SpringerBriefs in Computer Science

SpringerBriefs present concise summaries of cutting-edge research and practical applications across a wide spectrum of fields. Featuring compact volumes of 50 to 125 pages, the series covers a range of content from professional to academic.

Typical topics might include:

- A timely report of state-of-the art analytical techniques
- A bridge between new research results, as published in journal articles, and a contextual literature review
- A snapshot of a hot or emerging topic
- An in-depth case study or clinical example
- A presentation of core concepts that students must understand in order to make independent contributions

Briefs allow authors to present their ideas and readers to absorb them with minimal time investment. Briefs will be published as part of Springer's eBook collection, with millions of users worldwide. In addition, Briefs will be available for individual print and electronic purchase. Briefs are characterized by fast, global electronic dissemination, standard publishing contracts, easy-to-use manuscript preparation and formatting guidelines, and expedited production schedules. We aim for publication 8–12 weeks after acceptance. Both solicited and unsolicited manuscripts are considered for publication in this series.

**Indexing: This series is indexed in Scopus, Ei-Compendex, and zbMATH **

Zhiguo Shi • Chaojie Gu • Shibo He • Kang Hu

LoRa Localization

System Design and Performance Analysis

 Springer

Zhiguo Shi
College of Information Science and
Electronic Engineering
Zhejiang University
Hangzhou, China

Shibo He
College of Control Science and Engineering
Zhejiang University
Hangzhou, China

Chaojie Gu
College of Control Science and Engineering
Zhejiang University
Hangzhou, China

Kang Hu
IoT department
Alibaba Group
Hangzhou, China

ISSN 2191-5768   ISSN 2191-5776   (electronic)
SpringerBriefs in Computer Science
ISBN 978-3-031-48010-2   ISBN 978-3-031-48008-9   (eBook)
https://doi.org/10.1007/978-3-031-48008-9

This Springer imprint is published by the registered company Springer Nature Switzerland AG
The registered company address is: Gewerbestrasse 11, 6330 Cham, Switzerland

If disposing of this product, please recycle the paper.

Preface

The Internet of Things (IoT) has emerged as a transformative technology, enabling seamless connectivity and communication between various devices and objects. Within the realm of IoT, low-power solutions have gained significant attention due to their ability to support long battery life and enable widespread deployment of connected devices. One crucial aspect of low-power IoT is accurate node positioning, which holds immense potential for diverse industrial applications.

This monograph delves into the research and development of low-power IoT node positioning techniques, with a specific focus on the utilization of LoRa (Long Range) technology. LoRa offers long-range wireless connectivity, low power consumption, and cost-effective solutions, making it an ideal choice for IoT applications. Throughout this book, we explore different positioning models, hardware platforms, and algorithms to achieve accurate and efficient node localization in both indoor and outdoor environments.

Chapter 1 provides an introduction to low-power IoT, discussing its background, network architecture, and research on node positioning. Chapter 2 focuses on wireless positioning techniques and introduces a modular hardware platform for IoT applications. Chapter 3 explores wide area location using signal flight time and optimization techniques, achieving high accuracy with low power consumption. Chapter 4 presents a low-cost positioning system based on LoRa Mesh networking, demonstrating its effectiveness in wide-area coverage. Chapter 5 addresses indoor positioning challenges by utilizing signal arrival angles and antenna array structures, enabling accurate localization in complex environments. Chapter 6 investigates fusion localization and tracking using mobile robots, enhancing position estimation and trajectory tracking in diverse environments. Chapter 7 concludes the book, summarizing key findings and suggesting future research directions in LoRa's indoor and outdoor fusion positioning.

We hope this monograph serves as a valuable resource for researchers, engineers, and practitioners seeking to explore the potential of low-power IoT node positioning and contribute to the advancement of IoT technologies.

Hangzhou, China

Zhiguo Shi
Chaojie Gu
Shibo He
Kang Hu

Acknowledgements

We would like to thank all the people who have made contributions to this book. In particular, we want to acknowledge the enormous amount of help we received from Yihao Xu, Boya Liu, and Ying Liu at College of Information Science and Electronic Engineering, Zhejiang University; Haoran Shi at Polytechnic Institute, Zhejiang University; Yuhao Chen, Huimin Chen, and Jiming Chen at College of Control Science and Engineering, Zhejiang University; Tao Zhen at Alibaba Group.

In addition, we want to express our gratitude to our colleagues for their kind support and encouragement. Last but not least, this book is supported in part by the National Natural Science Foundation of China (NSFC) under grant U21A20456, U23A20326, 62302439, and U23A20296.

Contents

1 Introduction 1
   1.1 Background 1
        1.1.1 Internet of Things (IoT) 1
        1.1.2 Low Power Wide Area Network (LPWAN) 4
        1.1.3 Location-Based Internet of Things Services and
               Industrial Applications 5
   1.2 Current Situation and Challenge 8
        1.2.1 Challenges and Contributions 8
   1.3 Book Organization 9
   References 12

2 Wireless Localization Model and Hardware Foundation 13
   2.1 LoRa and Its Ranging Engine 13
        2.1.1 LoRa Network Structure 13
        2.1.2 LoRa Nodes Distance Estimation 14
        2.1.3 LoRa Nodes Angle Estimation 15
   2.2 Angle of Arrival (AoA) Estimation 16
   2.3 Hardware 16
        2.3.1 Hardware Platform Design 17
        2.3.2 Heterogeneous Devices and Protocol Interface Design 19
   Reference 20

3 LoRa-Based Mobile Localization System 21
   3.1 System Model 21
        3.1.1 System Overview 22
        3.1.2 Localization Model 22
   3.2 Distance and Location Estimation 24
        3.2.1 Distance Estimation 24
        3.2.2 Anchor Location Estimation 25
   3.3 Localization Optimization 26
   3.4 Implementation and Evaluation 30
        3.4.1 System Implementation 30

      3.4.2 Localization Simulations 31
      3.4.3 System Evaluation 33
  3.5 Summary 38
  References 38

4 **Wide-Area Localization System Based on LoRa Mesh** 41
  4.1 Hardware Design... 42
  4.2 LoRa Mesh Protocol Design.. 43
      4.2.1 The Structure of the Mesh Protocol 43
      4.2.2 Route Discovery and Maintenance................................ 45
      4.2.3 Routing Algorithm.. 46
  4.3 LoRa Ranging and Localization.. 47
      4.3.1 Localization Workflow 47
      4.3.2 Ranging Algorithm 49
      4.3.3 Localization Algorithm... 50
  4.4 Implementation... 51
      4.4.1 Control and Visualization Interface 51
      4.4.2 Anchors Deployment 52
  4.5 Evaluation 53
      4.5.1 Ranging Experiment... 54
      4.5.2 Positioning Experiment... 55
  4.6 Summary 58
  References 58

5 **Enable Angle of Arrival in LoRa for Efficient Indoor Localization** 59
  5.1 Problem Formulation and Challenges..................................... 59
  5.2 Redesign of LoRa Gateway and Ranging Procedure 60
      5.2.1 Gateway Redesign 60
      5.2.2 Leveraging Ranging Difference 61
      5.2.3 Ranging Procedure Redesign.................................... 62
  5.3 Improving AoA Estimation via Rotation 63
      5.3.1 Binary AoA Classification 64
      5.3.2 Virtual Array 65
      5.3.3 AoA Calculation 66
  5.4 Implementation and Applications 66
      5.4.1 Implementation 66
      5.4.2 Applications... 67
  5.5 Performance Evaluation 68
      5.5.1 Ranging Difference Estimation 68
      5.5.2 AoA Estimation... 69
      5.5.3 NLoS Localization.. 71
      5.5.4 Power Consumption Evaluation................................... 72
  5.6 Summary 73
  References 73

6 LoRa-Based Indoor Tracking System for Mobile Robots 75
   6.1 Estimating the AoA 75
       6.1.1 AoA Estimation with an Antenna Array...................... 76
       6.1.2 Minimize Ranging Interval.................................... 77
   6.2 Eliminating the Blind Area 79
   6.3 Estimating Target Movement 82
       6.3.1 Target Motion Model 82
       6.3.2 Real-Time Frequency Estimation 83
   6.4 System Implementation 85
   6.5 Performance Evaluation 86
       6.5.1 Indoor Experiments... 86
       6.5.2 Deploy Ability Investigation 91
   6.6 Summary 94
   References 94

7 Conclusion and Future Directions 95
   7.1 Concluding Remarks.. 95
   7.2 Future Directions... 96

Acronyms

3GPP	3rd-Generation Partnership Project
ADC	Analog-to-Digital Converter
AI	Artificial Intelligence
AoA	Angle of Arrival
BLE	Bluetooth Low Energy
CAD	Channel Activity Detection
CDF	Cumulative Distribution Function
CRC32	32-bit Cyclic Redundancy Check
CRLB	Cramer-Rao Lower Bound
COTS	Commercial Off-The-Shelf
CSI	Channel State Information
CWT	Continuous Wavelet Transform
DBPSK	Differentially Coherent Binary Phase Shift Keying
DTT	Data Transmission Time
eMTC	Enhanced Machine-Type Communication
FIM	Fisher Information Matrix
GFSK	Gauss Frequency Shift Keying
GNSS	Global Navigation Satellite System
GPS	Global Positioning System
ICT	Information and Communications Technology
IIC	Inter-Integrated Circuit
IMU	Inertial Measurement Unit
IoT	Internet of Things
ISM	Industrial Scientific Medical
LNA	Low Noise Amplifier
LoRa	Long Range
LoRaWAN	Long Range Wide Area Network
LoRaWAPS	LoRa Mesh-based Wide Area Positioning System
LOS	Line of Sight
LPWAN	Low Power Wide Area Network
LTE	Long-Term Evolution

MCU	Micro Controller Unit
MIMO	Multiple Input Multiple Output
MSE	Mean Squared Error
NB-IoT	Narrow Band Internet of Things
NLOS	Non-Line of Sight
OFDMA	Orthogonal Frequency Division Multiple Access
PDR	Pedestrian Dead Reckoning
PID	Proportional Integral Differential
PLC	Programmable Logic Controller
PRE	Posterior RSSI Error
RF	Radio Frequency
RSSI	Received Signal Strength Indication
SC-FDMA	Signal-Carrier Frequency Division Multiple Access
SGD	Stochastic Gradient Descent
SIM	Subscriber Identity Model
SLAM	Simultaneous Localization and Mapping
SMA	Sub Miniature-A
SNR	Signal-to-Noise Ratio
SPI	Serial Peripheral Interface
TDoF	Time Difference of Flight
ToF	Time of Flight
UNB	Ultra-Narrow Band
UWB	Ultra-Wide Band

List of Figures

Fig. 1.1 Hierarchical architecture of the Internet of Things 3
Fig. 1.2 Location-based IoT industry applications. (**a**) Intelligent transportation. (**b**) Logistics warehousing. (**c**) Intelligent agriculture. (**d**) Industrial manufacturing. (**e**) Intelligent city. (**f**) Intelligent medical treatment 6
Fig. 1.3 Outline 9
Fig. 2.1 LoRa network structure 14
Fig. 2.2 LoRa ranging process. (**a**) Ranging request. (**b**) Synchronization. (**c**) Ranging response 15
Fig. 2.3 LoRa hardware structure 16
Fig. 2.4 AOA estimation model 17
Fig. 2.5 Hardware platform structure 17
Fig. 2.6 Hardware physical design 20
Fig. 3.1 Architecture of iLoc 22
Fig. 3.2 Localization model 23
Fig. 3.3 Distance estimation 25
Fig. 3.4 Anchor location estimation 27
Fig. 3.5 Hardware and framework design. (**a**) LoRa tag. (**b**) Simplified mobile LoRa gateway. (**c**) Localization framework 32
Fig. 3.6 Optimized path in 2 km area. (**a**) $\lambda_1 = 0.30$. (**b**) $\lambda_1 = 0.35$. (**c**) $\lambda_1 = 0.40$ 33
Fig. 3.7 Optimized path in 400 m area. (**a**) $\lambda_1 = 0.30$. (**b**) $\lambda_1 = 0.35$. (**c**) $\lambda_1 = 0.40$ 33
Fig. 3.8 Error bound vs. path length 34
Fig. 3.9 CDF of localization error in different ranges 35
Fig. 3.10 CDF of localization error comparison 36
Fig. 3.11 CDF of localization error in different scenarios 37
Fig. 4.1 System architecture 42
Fig. 4.2 The hardware structure of the target 42
Fig. 4.3 The workflow of the mesh protocol 43

Fig. 4.4   The packet queue in physical layer 44
Fig. 4.5   The packet structure 45
Fig. 4.6   Flowchart of route lookup 46
Fig. 4.7   Positioning workflow 48
Fig. 4.8   Control and visualization interface 52
Fig. 4.9   Anchor deployment location selection and area coverage
             capability. (**a**) Alternative deployment locations.
             (**b**) Optimal placement of anchors. (**c**) Location capability
             coverage 53
Fig. 4.10   ToF performance test 54
Fig. 4.11   CDF of range algorithm 55
Fig. 4.12   CDF of localization algorithm 56
Fig. 4.13   Influence of anchor density on positioning ability 56
Fig. 4.14   Localization system test 57
Fig. 5.1   Antenna array model 61
Fig. 5.2   Ranging procedure redesign 62
Fig. 5.3   Adding "virtual antenna arrays" for AoA estimation.
             (**a**) Resolution of n uniform antenna. (**b**) AoA
             classification. (**c**) Rotate to localize. (**d**) IMU orientation
             estimation 65
Fig. 5.4   RLoc implementation that contains a tag and a redesigned
             gateway 67
Fig. 5.5   Experimental results of ranging difference 69
Fig. 5.6   Experimental results on AoA estimation 70
Fig. 5.7   Experimental results on AoA estimation 70
Fig. 5.8   Experimental results on AoA estimation 71
Fig. 5.9   Experimental results on NLoS localization in a long
             corridor 72
Fig. 6.1   We redesign the standard ranging procedure in 2.4 GHz
             LoRa. The antennas are selected by the controller via
             an RF switch, and the hopping procedure is optimized
             to reduce the time between TDoF estimations from
             two external antennas. (**a**) Antenna array structure.
             (**b**) Built-in hopping procedure. (**c**) Optimized hopping
             procedure 76
Fig. 6.2   The setup of the TDoF mircobenchmark. The target is put
             in the perpendicular bisector of the segment AB, making
             the ground-truth TDoF be 0 78
Fig. 6.3   Histogram with density curve of the TDoF measurement
             results 78
Fig. 6.4   Architecture model 79
Fig. 6.5   The basic idea of the virtual antenna array. In rotation, the
             two connected antennas emulate a circular antenna array 80

Fig. 6.6   The setup of the AoA estimation accuracy
            microbenchmark. We place the target at different
            locations separated by 5 degrees. The distance between
            the anchor and the target is 3 m 81

Fig. 6.7   To gain a better understanding, we translate the AoA into
            TDoF. In the blind areas, direct estimation with a static
            antenna array bias from the ground truth with an AoA
            error of more than 40 degrees 81

Fig. 6.8   The tracking model considers the tracking in a 2-D space.
            The anchor estimates the relative distance, angle, and
            velocity of the target 82

Fig. 6.9   Hardware implementation and firmware design of the
            LTrack system on a robot. (a) Hardware implementation.
            (b) Firmware framework 85

Fig. 6.10   LOS AoA estimation errors in different ranges 87

Fig. 6.11   LOS and NLOS AoA estimations 88

Fig. 6.12   NLOS AoA estimation error with blockage of different
            obstacles 50 m away from the anchor 88

Fig. 6.13   The tracking experiment in an indoor lab space. The robot
            begins tracking at (0, 0) and finishes at (3.8, 10) 89

Fig. 6.14   Obstacles in the lab and the tracking error. (a) Test bench.
            (b) Desk. (c) CDF of the tracking error 90

Fig. 6.15   Tracking experiment in a corridor. A person holds the tag
            in nature and walks at a speed of 0.3 m/s 91

Fig. 6.16   Tracking experiment in a corridor. A person holds the tag
            in nature and walks at a speed of 0.5 m/s 91

Fig. 6.17   Tracking errors with different walking speeds 92

Fig. 6.18   LTrack AoA estimation performance at different ranges 92

Fig. 6.19   Tracking simulation in a large indoor space with different
            safe distance settings 93

Fig. 6.20   Tracking errors with different safe distance settings 93

List of Tables

Table 3.1   Summary of notations 22
Table 3.2   Manufactory cost of four systems 34
Table 3.3   Power consumption comparison 38
Table 5.1   Experimental results of AoA estimation (unit: degree) 64
Table 5.2   Power consumption comparison 73

Chapter 1
Introduction

Abstract In this chapter, we first provide the background of Internet of Things, Low Power Wide Area Networks and location-based Internet of Things services. Based on the existing research work, we discuss the remaining problems and challenges related to indoor and outdoor localization and tracking of IoT nodes. Finally, we outline the scope and organization of this monograph.

Keywords Internet of Things · Low Power Wide Area Networks · LoRa · Location-based Internet of Things services

1.1 Background

1.1.1 Internet of Things (IoT)

With the rapid advancement of information technology, semiconductor technology, and communication technology, the focus of information networks has shifted from traditional computers and mobile networks towards the construction of a robust IoT that supports connectivity between everything. The IoT aims to enable seamless connection, perception, calculation, and control between people and devices, as well as devices and other devices. This provides an infrastructure for further development towards intelligent societies with capabilities for perception, analysis, and decision-making. In recent years, the number of IoT applications has exploded, playing an increasingly critical role in smart cities, transportation and material flow, energy and environmental protection, agriculture, rural construction, and other fields. According to statistics, as of 2021, the number of global IoT device accesses had reached 35.82 billion, about 4.5 times the world's population—and it is still in a period of high growth. It is estimated to reach over 75.54 billion access points by 2025 [1].

IoT has become a crucial strategy for countries to achieve leadership positions in science and technology development. In recent years, there has been global competition towards industrial transformation and upgrading with IoT at the core. The European Union (EU) was the first organization to propose development

© The Author(s), under exclusive license to Springer Nature Switzerland AG 2024
Z. Shi et al., *LoRa Localization*, SpringerBriefs in Computer Science,
https://doi.org/10.1007/978-3-031-48008-9_1

and management plans for the IoT. It has released policy documents outlining the IoT strategy and dedicated a significant investment through the Information and Communications Technology (ICT) research and development plan to support the project construction of IoT technology companies. The EU aims to lead the world in building the infrastructure necessary for IoT and usher in a new era of connected devices. The United States also possesses significant advantages in IoT development. The National Intelligence Council has identified IoT as one of the key strategic technologies in its report "Key Technologies of Potential Impact on American Interests in 2025". The US is thus pushing forward with its informatization strategic deployment to consolidate its position as a leader in information technology. Japan, a manufacturing powerhouse, has a highly advanced information technology and equipment manufacturing industry foundation. Since the start of the twenty-first century, Japan has implemented several information industry strategies such as "e-Japan strategy", "Active Japan ITC Strategy," and "Smart-Japan ICT Strategy" to create favorable conditions for the growth of its information industry.

The construction of IoT has become an integral part of China's strategic emerging industries. China is greatly committed to developing IoT and has made specific plans for its future growth. The 13th Five-Year Plan outlines that technological revolutions, such as IoT, will drive the evolution of cyberspace towards the "Internet of Everything," integrating the real world with the information world and promoting sustainable development and digital economy engines in service of the "Belt and Road" initiative. In developing IoT, China aims to profoundly reform the global governance system. It has put forth development requirements such as "accelerating the industrial agglomeration of IoT," "promoting entrepreneurship and innovation in IoT," and "accelerating the deep integration of IoT and industry." Additionally, it has planned key fields like "intelligent manufacturing," "intelligent agriculture," "intelligent home," "intelligent transportation and vehicle networks," "intelligent medical treatment," "health care for the elderly," and "intelligent energy saving and environmental protection." As China's economic construction enters the deep water zone, the "14th Five-Year Plan" emphasizes the construction of a digital economy, with IoT identified as the key industry. In particular, the plan highlights the importance of high-precision positioning technology innovation in the IoT industry as a key direction.

IoT can be divided into four distinct layers based on function and network level, including the sensing layer, network layer, platform layer, and application layer. As demonstrated in Fig. 1.1, these layers form a hierarchical structure. The data flow within the IoT begins at the sensing layer and passes through the network and platform layers before reaching the application layer. Conversely, the control flow starts at the application layer and then moves down through the platform and network layers to reach the physical device located in the sensing layer.

In the application of IoT, the sensing layer provides the terminal model and operation interface, while the network layer realizes the access of various devices and protocols. On this basis, the platform layer establishes a unified model and interface for heterogeneous devices and protocols, shields the differences of the

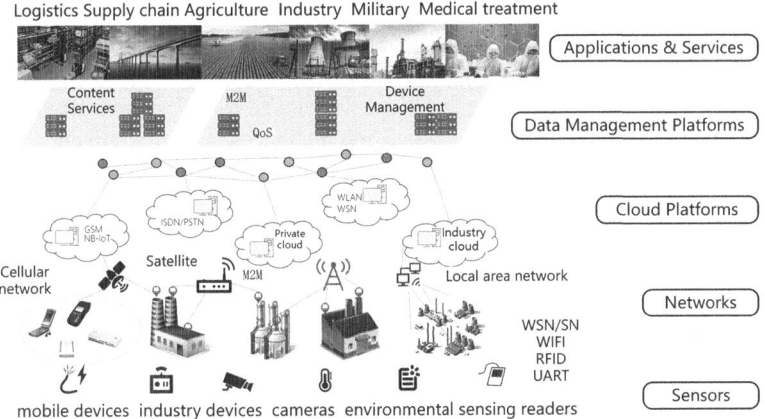

Fig. 1.1 Hierarchical architecture of the Internet of Things

bottom layer for the upper layer applications, and provides a unified application specification. Each layer has a different structure and equipment to achieve different functions.

1. Sensing layer: Sensing layer consists of all kinds of sensor nodes and controllers with certain control, calculation, and communication capabilities. It senses various states of temperature, pressure, speed, and current in equipment and environment through sensors. And the initial processing of the original data, with the implementation of the element to complete the corresponding action. The sensing layer is often embedded in the device side and can directly interact with the user. At the same time, the user can execute the operation instruction to realize the corresponding product service.
2. Network layer: Network layer is an important channel for data and instruction transmission, which is composed of various wired and wireless networks and communication protocols. The device transmits awareness information through the network layer and receives user commands from the application layer. The network layer includes information integrity checks, security encryption, and device access. In recent years, the emergence of low-power wide-area network technology has greatly improved the device capacity and coverage of the network layer, and provided the network foundation for the extensive application of IoT.
3. Platform layer: Platform layer is the key structure connecting devices and applications in the architecture of IoT. It is mainly responsible for unified management and security protection of all kinds of devices connected by the network layer, and is responsible for data processing, forwarding, storage, retrieval, and analysis. IoT platform provides a variety of applications in the application layer with reliable mass device management, data management, stable connection management, security management, and other services.
4. Application layer: The application layer obtains the node status and information from the platform layer, provides specific application services to users or devices,

and receives commands from users or devices to complete data and command interaction. It includes specific IoT applications for various industries, such as energy, transportation, industry, and agriculture, realizing a variety of industrial services. The application layer is typically deployed by IoT service providers and supports unified IoT services across different devices and environments.

1.1.2 Low Power Wide Area Network (LPWAN)

The network layer is responsible for node connection in IoT and incorporates numerous communication protocols tailored to different application scenarios. For instance, the 5G network, designed for high-speed image and video streaming data, can be utilized for autonomous driving, virtual reality, and other applications with high transmission rate requirements. On the other hand, WiFi, Bluetooth, and other networks, intended for medium and high-speed local device connections, can serve wearable devices, smart homes, and other applications that have low transmission distance and bandwidth demands. The Enhanced Machine-Type Communication (eMTC) network is fitting for wide area devices with moderate speed, appropriate for traffic and security applications with broad service distribution and specific transmission rate requirements.

Among all network technologies, medium and high-rate networks are mainly geared towards person-to-person connections. Conversely, connections between "things" rely on wide-area low-power networks with long distances and low power consumption as it pertains to the growth of IoT devices. Although the communication rates between them may not be high, their numbers are substantial, their distribution is wide-ranging, and both the communication distance and the operational life of the device require attention. Networking technologies for such IoT applications mainly comprise Long Range (LoRa), Sigfox, eMTC, and Narrow Band Internet of Things (NB-IoT).

LoRa is a physical layer protocol originally designed by Cycleo based on Chirp spread spectrum communication. It allows programmable data bandwidth, maintains sensitivity as low as −148 dBm, and enables urban coverage distances of 2–5 km with suburban coverage reaching up to 15 km. As one of the first commercially available low-power wide-area networks, LoRa works on unlicensed Industrial Scientific Medical (ISM) bands and has the backing of many countries and regions. To promote the growth and application of LoRa, several global telecom operators, network service units, equipment production units, and application development units formed an open non-profit LoRa alliance organization. They launched a standard media access control layer protocol called Long Rang Wide Area Network (LoRaWAN), which provides unified access standards.

Significantly, like LoRa, SigFox is a proprietary networking technology operating on the unlicensed ISM spectrum. It comprises a complete physical layer, media access control layer, and application layer architecture. Unlike LoRa, it uses ultra-narrow band (UNB) modulation technology to transmit short messages. This translates to each message containing at most 12 bytes with a maximum of 144 messages processed per day [2]. Sigfox frequency bands are less than 1 GHz, with Europe using 868 MHz while the US employs 915 MHz. Unidirectional or half-duplex bidirectional communication occurs between nodes and base stations. For downlink transmission, Differentially Coherent Binary Phase Shift Keying (DBPSK) modulation is used whereas Gauss Frequency Shift Keying (GFSK) modulation supports uplink transmission. Additionally, the Sigfox link budget extends up to 140 dB, enabling the cooperative reception of node signals by multiple base stations resulting in ultra-long-range communication range for suburbs of 30–50 km [3].

eMTC is a new IoT technology based on Long Term Evolution (LTE) evolution that supports voice data transmission and boasts a rate of nearly 1 Mbps. This network was first proposed by 3rd Generation Partnership Project (3GPP) in the R13 version as the initial standard for machine communication. Orthogonal Frequency Division Multiple Access (OFDMA) technology is employed for the uplink, while Signal-Carrier Frequency Division Multiple Access (SC-FDMA) technology is utilized for the downlink. Compared to the LTE protocol, eMTC has been optimized and tailored to reduce system costs. Unlike LoRa and Sigfox technologies, eMTC works within authorized frequency bands and uses cellular networks for successful deployment. Remarkably, every single cell in this technology can support the connection of 10,000 nodes located by the base station by the signal arrival time difference [4].

NB-IoT is an international standard proposed by 3GPP beginning in 2016 Release 13. Similar to eMTC, the NB-IoT network uses commercial licensed frequency bands, and device terminals connect to the network via carrier base stations. The downstream and upstream transmission schemes are OFDMA and SC-FDMA, respectively. NB-IoT nodes utilize the LTE discontinuous receiving mechanism to reduce terminal power consumption by extending the discontinuous receiving mode to adjust the paging period length or turning off the network paging mode to enter the power-saving mode. However, unlike the eMTC protocol, NB-IoT's data rate is low with upstream rates ranging from approximately 160 to 200 kbps and downstream rates from about 160 to 250 kbps. This makes it more suitable for low-speed and low-power wide area IoT applications.

LPWAN has clear advantages in terms of network capacity, coverage distance, and power consumption. It provides technical support for IoT applications on a massive scale. When compared to other technologies, LoRa boasts a more open and flexible network structure, lower manufacturing costs, and a longer service life. Consequently, it has been widely used in various regions and fields across the world.

1.1.3 Location-Based Internet of Things Services and Industrial Applications

In light of the growing variety and quantity of IoT applications, as well as the increasing scale of nodes that utilize such technology, node location information is gradually becoming a crucial factor in the optimization of network performance and node management. Through an analysis of the usage of location information, it is possible to categorize location-based services in IoT applications into two distinct groups: namely, location-based applications and location-based optimization.

Location-based application: In an array of emerging application scenarios, positioning, navigation, and management services can be provided through a system that is based on node locations; as seen in the example of animals located in intelligent pasture, packages situated within intelligent logistics and storage, and target objects located during disaster relief efforts. In these cases, the location of the terminal node is directly presented to the user who can monitor or control said node through the use of location information.

Location-based optimization: On the other hand, in large-scale IoT applications that feature a significant number of nodes dispersed over vast areas, the system can optimize communication, energy consumption, and other resources while simultaneously coordinating tasks via node location information. For example, unmanned ports may employ multiple robots working together based on position status information, while traffic flow scheduling and road management are improved through the use of location-based information within the traffic system. By utilizing location as principal data for optimization, the system can increase management efficiency and decrease overall resource consumption.

Numerous industrial applications that rely on location-based services play a critical role in various fields, including medical care, transportation, logistics, and urban development. These applications have significant implications for economic development and social progress. As illustrated in Fig. 1.2, the industrial application of IoT that employs location-based services primarily comprises the following aspects:

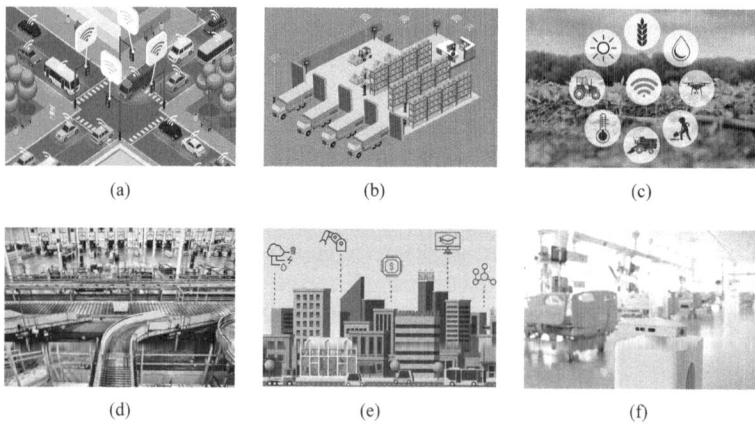

(a)   (b)   (c)

(d)   (e)   (f)

Fig. 1.2 Location-based IoT industry applications. (**a**) Intelligent transportation. (**b**) Logistics warehousing. (**c**) Intelligent agriculture. (**d**) Industrial manufacturing. (**e**) Intelligent city. (**f**) Intelligent medical treatment

The transportation and logistics systems serve as the lifeline of economic and social development. In this field, spatiotemporal information pertaining to targets is continually changing, and obtaining vehicle position status via network and location service enables real-time road traffic dispatching and enhances transport capacity. By relying on location information for task coordination and optimization of logistics robots in ports and warehouses, it becomes feasible to accomplish status perception, real-time analysis, and effective management of traffic and logistics systems. By doing so, the intelligent level of traffic and logistics services can be significantly improved.

In the realm of industrial manufacturing, the advent of emerging technologies such as IoT, cloud computing, and big data has propelled the global manufacturing industry towards a new phase of the industrial revolution. During the transformation and modernization of this industry, real-time perception of the position and state of human, machine, material, and material serves as a vital technology to facilitate the shift towards networking, refinement, and intelligence in production, manufacturing, and management. By means of real-time collection and analysis of production status information, efficiency, and quality can be enhanced while simultaneously advancing safety, energy conservation, and emission reduction. As such, this technology enables the promotion of sustainable development and the transformation and modernization of the manufacturing industry.

In the domain of urban construction and management, IoT technology enables the achievement of mutual perception and interconnection between people and things and things. This technology facilitates the acquisition of spatiotemporal information of various targets within a city through location services, thereby promoting urban information sharing, complementary advantages, and collaborative development. The utilization of this technology enhances the efficiency of urban management, particularly in major social events such as disaster relief and epidemic prevention and control. By analyzing the spatial and temporal correlation based on comprehensive location information of the target population, the ability to govern urban areas and provide public services can be augmented.

Medical care and elderly care are two critical aspects related to the well-being of individuals. In contemporary times, IoT technology, utilizing location information, has assumed significant importance in both these domains. By employing sensing and positioning technology, online monitoring and tracking of life science observation objects can be achieved, ensuring safe and reliable monitoring services by mobile robots. Furthermore, this IoT technology facilitates the quick tracking, positioning, and tracing of medical materials such as drugs and medical waste, thereby reducing supervision costs and improving the quality of medical and elderly care services.

1.2   Current Situation and Challenge

1.2.1   *Challenges and Contributions*

After reviewing pertinent literature, the research gaps and challenges in the integration of indoor and outdoor positioning and tracking of IoT nodes are identified as follows:

Node power consumption and cost: When implementing large-scale IoT applications, the power consumption and cost of battery-powered nodes become crucial considerations. Complex positioning algorithms and expensive sensors can severely impact system resources and overall costs. Therefore, developing efficient positioning algorithms and simplifying hardware structures is essential for reducing power consumption and hardware costs during node positioning.

Network coverage and deployment: In the case of widely distributed IoT nodes, the positioning system must ensure complete network coverage. One significant challenge is to determine how to use long-distance communication capabilities of low-power wide-area networks to estimate a node's position. Additionally, deploying an effective system across a wide area using LoRa networks that operate in unlicensed frequency bands requires consideration of various factors, such as realizing node positioning, avoiding dependence on a large number of network base station infrastructures, and improving system deployability.

Estimation of signal arrival Angle in resource-constrained hardware: In resource-constrained hardware, estimating signal arrival angle poses a significant challenge for simplified IoT nodes. These nodes have limited hardware resources such as small computing power and bandwidth, and usually only contain a single Radio Frequency (RF) channel, while the estimation of signal arrival Angle depends on the antenna array structure. Achieving effective signal arrival angle estimation on nodes with limited hardware resources is crucial for node localization.

Positioning accuracy in complex indoor and outdoor scenes: Positioning accuracy in complex indoor and outdoor scenes presents a challenge due to the wide range of applications possible with IoT covering various indoor and outdoor scenarios with complex differences. Ensuring the stable positioning of nodes in different scenarios and overcoming environmental interference to meet varied application requirements is challenging.

Trajectory estimation for moving targets: Real-time estimation of the trajectory of moving targets for node tracking is necessary for many mobile applications. Designing a simple and efficient gateway structure based solely on LoRa signal characteristics of nodes, realizing target motion state estimation, and developing an efficient algorithm to complete online trajectory calculation represent a significant challenge.

1.3 Book Organization

The positioning of IoT nodes in various indoor and outdoor scenarios has become an important area of interest for both academia and industry due to its vital role as a fundamental supporting technology and service in numerous IoT applications. Unlike traditional positioning methods, IoT node positioning must take into account the unique characteristics of low power consumption, cost-effectiveness, wide network coverage, and compatibility with various indoor and outdoor scenes. Existing positioning technologies have struggled to meet the diverse needs of IoT applications effectively.

To address this practical need, the book proposes LoRa localization and tracking technology, which offers full scene coverage. The proposed outline is depicted in Fig. 1.3.

Chapter 1 serves as an introduction to this monograph. It initially discusses the research background and significance of the low-power IoT, highlighting its typical network architecture, main technical characteristics, and diverse industrial applications. Drawing upon relevant cutting-edge research works and achievements, this chapter classifies and summarizes existing research on node positioning while identifying the associated deficiencies and challenges. It also outlines the book's main research content and organizational structure.

Chapter 2 mainly focuses on introducing the wireless positioning model of IoT nodes and the foundation of the IoT hardware platform. Building upon the primary positioning techniques summarized in Chap. 1, this chapter first analyzes the positioning method based on signal distribution characteristics and delineates its advantages and limitations concerning accuracy and cost. Furthermore, the principle of time-of-flight positioning based on signals is introduced with a detailed analysis of the primary factors limiting positioning accuracy within the model. The benefits of using this model in wide-area scenarios are also discussed. To support complex

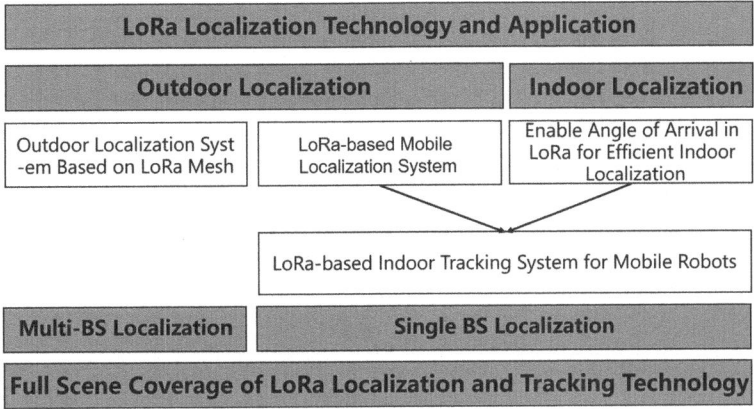

Fig. 1.3 Outline

indoor scenes, the positioning principle based on the signal angle is analyzed, and different estimation methods for various signal angles are compared, with further analysis of the accuracy performance of the angle positioning model. This chapter provides a theoretical foundation for subsequent chapters. Additionally, it introduces a modular general hardware platform with layered hardware and software that integrates heterogeneous protocols and device interfaces, which provides a hardware foundation for the following chapters' research on IoT application and positioning.

Chapter 3 introduces the wide area location of a single base station based on signal flight time. In an outdoor wide-area scenario, nodes require good network coverage and low energy consumption. Low power and high sensitivity LoRa networks allow nodes to estimate their positions based on observations from multiple base stations. However, in practical applications, the environment can limit base station deployment, resulting in increased system costs with numerous base stations.

This chapter designs a node location algorithm of a single base station based on the basic principle of multilateral positioning. Specifically, a single gateway is time-multiplexed to form a group of virtual base stations that locate nodes through mobile mode. The gateway's moving trajectory is optimized based on the Crmeralo-bound's optimal location estimation, and Inertial Measurement Unit (IMU), magnetometer, and Global Positioning System (GPS) are integrated to improve node positioning accuracy and reduce power consumption. A portable gateway structure, a card-type location tag, and a location application deployable on commercial mobile phones are designed, and the algorithm's performance is verified in various scenarios.

Experimental results demonstrate that the system achieves an average positioning error of 1.19 m in an outdoor open environment of 10,000 square meters, requiring only 1.06 mAh of power consumption.

Chapter 4 presents a study on outdoor wide-area localization based on LoRa Mesh. Many positioning tasks in an outdoor, wide-area environment require node devices to meet conditions of low power consumption, long endurance, and low cost, while also requiring the positioning system to achieve high-precision positioning and wide-area coverage.

This chapter designs a low-cost and low-power outdoor positioning system based on multi-anchor wireless Mesh networking and multi-dimensional data fusion using LoRa technology. Based on the designed LoRa Mesh protocol, ranging and positioning algorithm, the positioning system achieves the coexistence of positioning and communication functions at the system level by optimizing the system logic of networking communication and wireless positioning functions.

Test results conducted in the campus environment demonstrate that this system can provide good location services for the campus area. The peak power consumption of a single device in the system is less than 120 mW and the cost is less than $10, meeting the outdoor positioning requirements of low-power and low-cost devices.

Chapter 5 delves into the study of indoor positioning based on LoRa signal arrival angle. In complex indoor environments, the flight time and space distance of signals after reflection no longer satisfy a linear relationship, resulting in the failure of the multilateral positioning model. To address this limitation, this chapter presents a localization method based on signal arrival angle. However, measuring signal angle poses challenges in low power wide area networks due to hardware limitations.

To overcome this obstacle, this chapter designs an Angle estimation method based on signal flight time difference and establishes a spatial model between signal flight time difference and signal arrival Angle, enabling LoRa-based signal arrival Angle estimation. The chapter first proposes an RF channel multiplexing structure that extends a single channel into multiple channels to form an antenna array structure by connecting multiple antennas. Moreover, the physical layer protocol is simplified, and a pipeline algorithm of channel switching is designed, reducing the sampling interval time of the signal and suppressing environmental noise.

Furthermore, a large-scale circular array is simulated by rotating the gateway antenna combined with the spatial attitude of the gateway to improve the accuracy of Angle estimation. This chapter also designs the prototype of a rotating gateway and node and carries out experiments and simulations based on this prototype. The experimental results demonstrate that when the indoor distance is 50 m, the average error of the system signal arrival Angle estimation is 5.55°, enabling the realization of a single base station node location in complex indoor environments.

Chapter 6 investigates fusion localization and tracking based on mobile robots. The integration of indoor and outdoor positioning technologies by utilizing the variance criterion is studied for IoT nodes that are widely distributed both indoors and outdoors. The node positioning model parameters are dynamically adjusted based on the distance position estimation variance and Angle position estimation variance to overcome complex environmental noise, thereby improving the accuracy of node position estimation.

To address software and hardware resource constraints in the trajectory tracking of moving nodes amidst complex environmental interference, the time-frequency characteristics of the LoRa signal were analyzed, and the target speed was calculated based on the Doppler frequency shift model. The robot's moving path was optimized while considering the environmental characteristics measured by the robot, and the robot was guided to follow the moving target. Experiments and simulations based on commercial robots in various complex indoor environments show that the average fusion positioning error of the system is 0.92 m. In the trajectory estimation of a 600 square meters indoor corridor and a 3500 square meters outdoor square, the average error is 0.60 m and 0.33 m, respectively, with corresponding standard deviations of 0.49 m and 0.22 m, respectively. Moreover, this method can track multiple targets simultaneously.

Chapter 7 provides a summary of the book and outlines directions for future research on LoRa's indoor and outdoor fusion positioning, as well as open questions requiring further investigation.

References

1. The ultimate list of internet of things statistics for 2022[R]. Technical report, Statistics (2022)
2. B.S. Chaudhari, M. Zennaro, *LPWAN Technologies for IoT and M2M Applications[M]* (Academic Press, Cambridge, 2020)
3. K. Mekki, E. Bajic, F. Chaxel, F. Meyer, Overview of cellular LPWAN technologies for IoT deployment: Sigfox, LoRaWAN, and NB-IoT[C], in *Proceedings of the 2018 IEEE International Conference on Pervasive Computing and Communications Workshops* (2018), pp. 197–202
4. M. El Soussi, P. Zand, F. Pasveer, G. Dolmans, Evaluating the performance of eMTC and NB-IoT for smart city applications[C], in *Proceedings of the 2018 IEEE International Conference on Communications* (2018), pp. 1–7

Chapter 2
Wireless Localization Model and Hardware Foundation

Abstract In this chapter, we first introduce the basic principle of localization. Based on this, we analyze the distance-based localization model and angle-based localization model. Finally, we design a common hardware platform for various heterogeneous devices and protocols in IoT application and localization research.

Keywords Localization model · LoRa · Distance estimation · Angle estimation · Hardware platform

2.1 LoRa and Its Ranging Engine

LoRa, for its low power consumption and long-distance communication, owns obvious advantages in IoT localization applications. This section takes LoRa localization as an example to introduce LoRa network structure, distance estimation process, and angle estimation principle.

2.1.1 LoRa Network Structure

LoRa, for the advantage of flexible network structure and low-cost hardware devices, becomes one of the most popular IoT technologies, recognized globally by countries and regions. As a physical layer protocol based on spread spectrum communication, LoRa works in ISM band. Therefore, commercial LoRa devices can be applied to construct flexible application networks. LoRa Alliance, consisting of enterprises and organizations, has released LoRa media access control layer standard named LoRaWAN [1], so as to unify LoRa network standard. As shown in Fig. 2.1, LoRaWAN defines a star network structure with terminal nodes and gateway devices. Half-duplex communication is used between terminal nodes and gateways to upload node data and download gateway instructions. The gateway connects to the IoT cloud platform or user data server by means of the IP network. In order to apply to different applications and make full use of network resources,

Z. Shi et al., *LoRa Localization*, SpringerBriefs in Computer Science,
https://doi.org/10.1007/978-3-031-48008-9_2

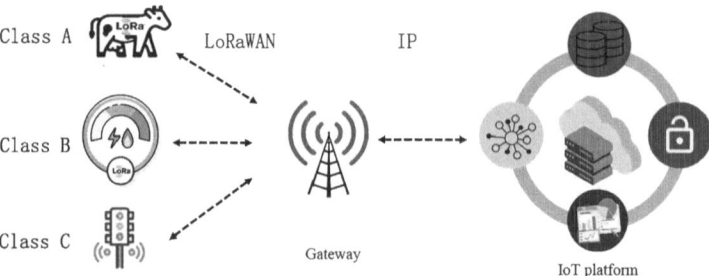

Fig. 2.1 LoRa network structure

LoRaWAN divides nodes into three categories: A, B, and C. Nodes A apply to upstream data applications with ultra-low power consumption. The nodes only send messages to the gateway randomly when sending data and then monitor downlink messages of the gateway for two receiving-time windows. Finally, the nodes turn to sleep mode and keep the mode in other conditions, unable to be waked up in the asynchronous state with the gateway. For the downlink user message at any time, the gateway sends it to the nodes in the receiving window only after receiving the data from the nodes. The nodes obtain obvious advantages of power consumption for the long-time sleep state. Based on the working mode of node A, node B sends uplink data, also maintaining two receiving windows. Differently, node B maintains time synchronization with the gateway through the downlink synchronization beacon of the gateway. Therefore, node B will periodically open the synchronous receiving window to receive synchronization signals, and adopt a dynamic receiving window width to suppress the influence of time drift. Nodes B can open the data receiving window to obtain downlink data at a specified period after achieving the synchronization with the gateway. Node B only opens the receiving window when data is sent or within the specified period. They are also in sleep mode at other times, achieving a balance between low power consumption and data transmission. For real-time downlink applications, nodes C, based on nodes A, maintain the receiving state to ensure the timely arrival of downlink messages. Therefore, synchronization with the gateway is not required. In conclusion, node types can be flexibly converted in terminal devices to adapt to different working modes in specific applications.

2.1.2   LoRa Nodes Distance Estimation

To support location-based IoT applications, the 2.4 GHz LoRa hardware is integrated with ranging engine to measure the Time of Flight (ToF) of signals between nodes, and estimates the distance. As shown in Fig. 2.2, the physical layer addresses of the host node and slave node are known during the ranging process. First, the host node sends the ranging request frame, containing physical layer address, check

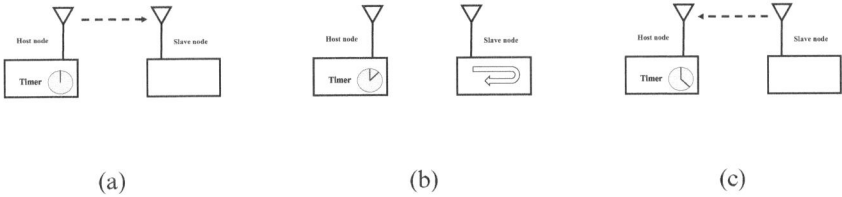

Fig. 2.2 LoRa ranging process. (**a**) Ranging request. (**b**) Synchronization. (**c**) Ranging response

code, ranging mark, and so on. The host node starts an internal timer after sending the ranging request. When the ranging request reaches the slave node, the slave node first checks whether the destination address is corresponding to the sending address. Once the addresses are corresponding, the ranging request is synchronized and the ranging response frame is sent to the host node. The synchronized time is determined in the host node. When the ranging response frame is returned to the host node, the host node checks the frame address and stops counting.

The host node is responsible for timing and output signals for two-way flight time during the process of ranging. The slave node and the host node sample different clock sources respectively. In the process of dealing with the ranging frame after receiving the ranging request, the clock will fluctuate with the temperature and other environmental factors due to crystal vibration. In this case, the fixed synchronization time between the host node and the slave node is not strictly corresponding, leading to the time difference. we can exchange the ranging process between the slave node and the host node to eliminate the time difference based on Model (2.1), where t_e represents the deviation of time fluctuation and t_{rtof} represents bidirectional signal time of flight.

$$t_{tof}^{ij} = \frac{(t_{rtof}^{ij} + t_e) + (t_{rtof}^{ji} - t_e)}{2} \tag{2.1}$$

The ToF precision of LoRa nodes is corresponding to the communication band. In wireless ranging systems, we define system time resolution as $\frac{1}{B}$, B is the signal bandwidth. Different from other narrow-band communication protocols, LoRa can improve its signal bandwidth by means of spread spectrum, which improves the ranging time resolution up to $\frac{1}{2^{12}B}$ in certain modes.

2.1.3 LoRa Nodes Angle Estimation

The analog front end of the communication system in commercial LoRa devices includes a low-noise amplifier, power amplifier, analog-to-digital conversion, and so on. The modem, power supply, and data transfer modules are integrated into a single chip, as shown in Fig. 2.3. Due to hardware cost and power consumption, the system contains only a single RF channel and no measurement unit that provides a

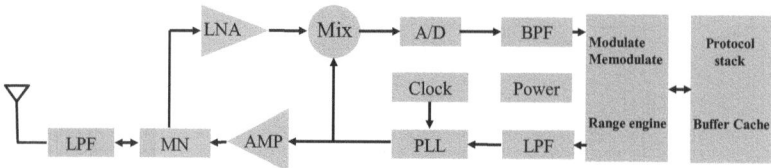

Fig. 2.3 LoRa hardware structure

signal phase. In the process of signal arrival angle estimation, the target node signal can be collected by the external antenna array and we can estimate by means of spatial spectrum. It has high requirements in terms of system power consumption and cost convenience for the reason that the structure requires multiple receiving devices to sample signals simultaneously.

2.2 Angle of Arrival (AoA) Estimation

Many Multiple Input Multiple Output (MIMO) devices incorporate array antenna structures, such as WiFi and network cards, to analyze channel state information such as signal phase. AoA estimates the target node through the angle at which the target node transmits or reflects signals to the antenna array and the position of the anchor point. In this case, the anchor point independently observes the angle information of the target node without time synchronization. In the 2D plane, considering the anchor points $A_i, i = 1, 2, ..., m$, estimate the angle of target nodes $T_j, j = 1, 2, ..., n$, the result shows as $\tilde{a}_{ij}$.

As shown in Fig. 2.4, the coordinate and angle between the anchor point and the target node satisfy the relation $\tan \alpha = \frac{y_j - y_i}{x_j - x_i}$. When the anchor point completes the signal arrival angle estimation of the target node, we calculate the target node position $\hat{T}$ by minimizing the difference between the observed signal arrival angle and the real angle. The specific formula is as follows:

$$\hat{T} = \arg \min_T \sum_{i=1}^{m} \sum_{j=1}^{n} \left\| \tilde{d}_{ij} - d_{ij} \right\| \tag{2.2}$$

2.3 Hardware

In the research of IoT applications and localization, specific hardware platforms are required for different positioning methods, protocols, and devices. Considering the complex and diverse types of sensors and networks, designing specific hard-

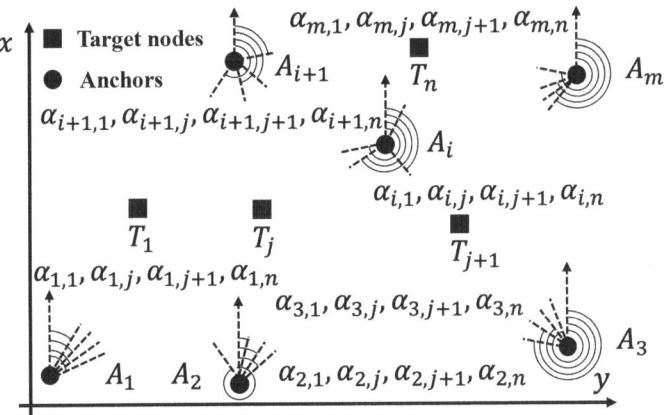

Fig. 2.4 AOA estimation model

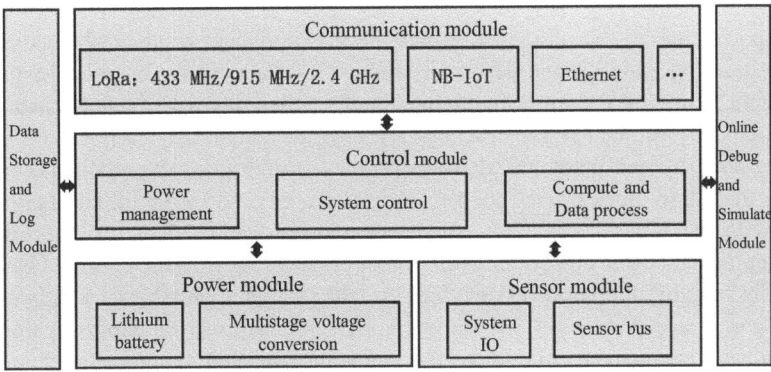

Fig. 2.5 Hardware platform structure

ware systems in different scenarios increases the time and cost of research and application. In this section, combined with the characteristics of heterogeneous devices and protocols, we design a general hardware platform supporting a variety of networks and interfaces, providing the hardware foundation for IoT applications and localization research.

2.3.1 Hardware Platform Design

In this section, to achieve rich product services in various IoT applications, a universal hardware platform is designed to support IoT applications, design, and validation of rapid localization research, combined with different types of devices and network characteristics. The hardware platform structure shows in Fig. 2.5, which can be divided into a power module, control module, communication module, sensor module, data storage module, and system simulation module. STM32L476 low-power Micro Controller Unit (MCU) is the controller in the hardware platform, connecting the communication module by means of Serial Peripheral Interface (SPI) bus. The communication module contains LoRa chip SX1276 in sub-1 GHz, LoRa chip SX1280 in 2.4 GHz for ranging, BC26 module for NB-IoT, and W5500 chip for Ethernet. Kinds of sensor interfaces are integrated into the hardware platform for sensor connection in different scenarios and the sensor data is saved in a data storage module by means of control module. In the process of application design and validation, the ST-Link simulation module can be applied for system debugging online. The universal hardware platform provides a rounded evaluation and validation system for various of IoT nodes and the design and optimization in the aspect of the network, power consumption, and so on can be further researched.

2.3.1.1   Low-Power Supply Design

The power module is a significant part of the hardware system, which directly influences the stability and pow consumption of the system. In order to satisfy different voltage and current requirements of various devices in IoT applications, hierarchical-multipath power structures are integrated, achieving the goals of system power stability and hardware platform power consumption. In particular, considering that many devices are independently powered, 2680 mAh lithium battery is added to the hardware platform to power independently. Moreover, the lithium battery management chip IP5306 can be applied in the management of charging and discharging. In the process of charging, the system depresses the supply of the type-c interface from 5 to 4.2 V and sends charging states to the controller through the power management chip. In the process of discharging, the power module rises the power from 4.2 to 5 V through the DC-DC module, providing power for system peripherals. For the 3.3 V voltage devices like the controller and LoRa chip, the platform contains the structure of multi-power, providing independent power for the controller, LoRa module, NB-IoT module, and Ethernet module through RT8509 chip. Moreover, the platform can configure the output state of the power to enable the power of the related system to reduce the system power consumption under the corresponding working conditions. Only the controller power is reserved on the hardware platform when the system is in hibernation mode, when the system electric current consumption is low to 5 μA, satisfying the requirement of long-time work in the conditions of lithium battery power.

2.3.1.2   High Precision Clock Design

Different modules in the hardware platform have different clock frequencies. Each module designs an independent clock system for management, specifically including LoRa high precision clock in 2.4 GHz, LoRa clock in sub-1 GHz, ethernet clock, and controller high speed clock and low speed clock. IQXT-205 oscillator is implied in the 2.4 GHz LoRa clock for ranging and communication clock source.

The precision of the crystal clock is 1.5 ppm and possesses low phase noise. Based on the clock, the system generates the carrier frequency corresponding to the 2.4 GHz band through the phase locked loop, providing a stable clock signal for ranging. The system clock is necessary to support controller working in different power consumption modes. Therefore, the controller system clock, including high frequency crystal in 8 MHz and low frequency crystal in 32.768 kHz, is equipped in the system. The high frequency clock reaches 80 MHz after frequency doubling, providing the clock signal in high performance mode. While in sleep mode, the high frequency clock stops working to reduce power consumption, and only the low frequency clock maintains the system. For Ethernet module, NB-IoT module, and low frequency LoRa module, the clock is provided by separate crystals, configured respectively in different applications.

2.3.2   Heterogeneous Devices and Protocol Interface Design

In the IoT applications and localization research, this section mainly designs abundant interfaces from the aspect of sensor and network transmission to support heterogeneous devices and protocols and realizes the hardware foundation support of several application scenarios based on open general software and hardware interfaces and services.

2.3.2.1   Sensor Bus Interface Design

In different application scenarios, the information sensed by node terminals differs greatly from the sensor types. The sensor interface can be divided into serial port, SPI, Inter-Integrated Circuit (IIC), and Analog-to-Digital Converter (ADC) by means of connection and communication between various sensors and controllers. We design two serial ports on the platform to match different kinds of heterogeneous devices, one of which supports differential signal RS485 to access the Programmable Logic Controller (PLC) and other host equipment. Moreover, we preserve two SPI buses, three IIC buses, and several ADC ports, as shown in Fig. 2.6. The hardware platform obtains the status of external devices through the bus or analog interface and sends data instructions. Since the implementation of interface protocol is closely related to the configuration of system hardware resources, a unified sensor interface device is designed in the platform to manage various external devices in the system away from hardware resource conflicts and mismatches. In particular, the hardware driver layer provides data read and write and parameter setting methods for different access protocols. Meanwhile, the state lock is set to manage communication interface objects during the process of hardware resource access. It will suspend data read and write requests when the bus is occupied and return communication resources when the bus is idle in order to reduce data read and write conflicts.

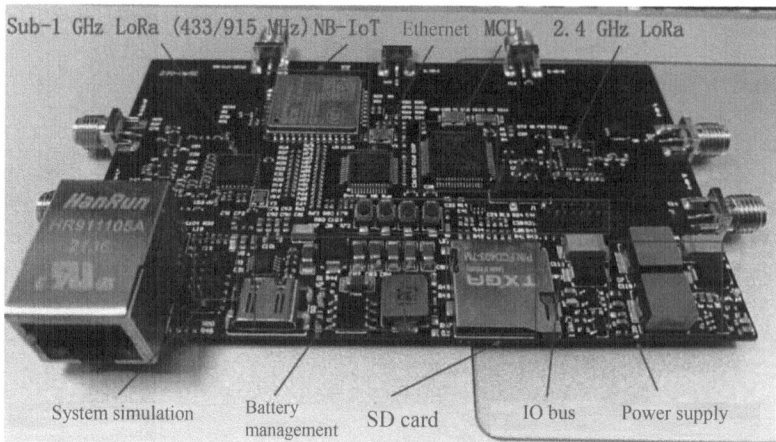

Fig. 2.6 Hardware physical design

2.3.2.2   Heterogeneous Network Protocol Interface Design

In regard to different protocols in IoT applications, LoRa in sub-1 GHz, LoRa in 2.4 GHz, NB-IoT, and Ethernet are mainly integrated into the hardware platform. SX1276 is the low-frequency LoRa chip, supporting sending and receiving signals in 433 and 915 MHz. SX1280 is the high-frequency LoRa chip in 2.4 GHz, supporting communicating and ranging between nodes. To satisfy the need for localization applications and research, the single RF channel in SX1280 is extended to four Sub Miniature-A (SMA) ports by means of a single-pole and four-throw RF switch HMC241 to connect the external antenna, realizing the measurement of the difference between the signal flight time. BC26 module and Subscriber Identity Model (SIM) card interface are equipped in the hardware platform for the NB-IoT to connect directly to the commercial network base station. The BC26 module communicates with the microcontroller through a serial port while the internal module sets OpenMCU open to support the research and design of the network and applications. LoRa and NB-IoT terminals are also designed, providing the hardware foundation for the fusion and evaluation of the low-power and wide-area heterogeneous network. The hardware platform supports TCP/IP protocol for the reason of the integration of Ethernet chip W5500, which can be applied as a gateway in IoT system to connect the platform and nodes, forming a complete hardware system structure. The hardware platform implements respectively network communication interface firmware to support device connection and data transmission for LoRa, NB-IoT, and Ethernet, providing abundant network interface for IoT applications and localization research.

Reference

1. LoRaWAN, Accessed: 2022 [Online]. Available: https://lora-alliance.org/about-lorawan/

Chapter 3
LoRa-Based Mobile Localization System

Abstract The node location has been regarded as one of the most important pieces of information to many novel applications of Internet of Things (IoT) in recent years. Typically, IoT nodes are always resource-constrained in terms of battery capacity, manufacturing cost, and computing ability. Thus, costly and energy-hungry localization techniques fall short for those applications. In this chapter, we present iLoc, a low-cost, low-power, and wide-area mobile localization system for IoT applications. iLoc is built on the emerging LoRa technologies and overcomes most disadvantages of the local localization techniques. Central to iLoc is a mobile anchor comprised of a simplified gateway and a commercial smartphone. To locate an IoT target node, the mobile anchor moves in a scheduled way, during which the gateway receives its locations from the smartphone and measures the time of flight (ToF) from the target node as well. The measurements are jointly optimized via fusing the RSSI and ToF in the ranging. Meanwhile, the inertial measurement unit, the GPS receiver, and the magnetometer are also fused in the anchor location estimation. We further design an iterative localization algorithm by judiciously deciding the locations of the anchor in an optimized way that not only minimizes the localization error bound but also reduces the system power consumption. We prototype the gateway and IoT node that cost less than 10 and 5 dollars, respectively. The extensive experiments demonstrate that iLoc achieves a mean localization error of 1.19 m and the power consumption is less than 1.06 mAh in an open outdoor environment.

Keywords Location-based services · Internet of Things · LoRa · Low-power wide-area networks · Time of flight

3.1 System Model

In this section, we first introduce the system overview, symbol notations, and then elaborate on the localization model.

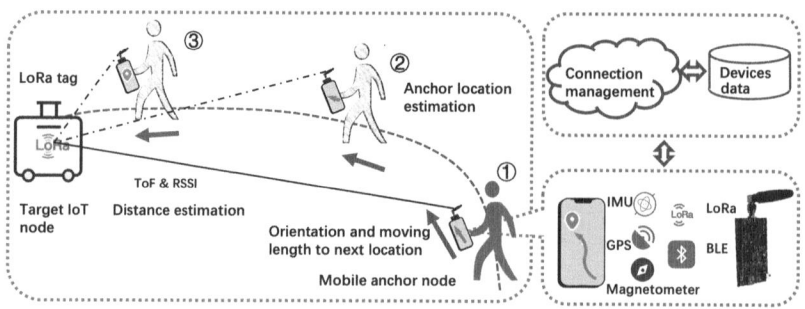

Fig. 3.1 Architecture of iLoc

Table 3.1 Summary of notations

Symbol	Definition
s	Target node location that need to be estimated
$\mathbf{r}_i$	Anchor node location at i-th step
$\tilde{\mathbf{r}}_i$	Anchor node's measurement locations at i-th step
$\mathbf{r}$	Anchor node's locations vector during the moving
d_i	Distance between anchor and target node at i-th step
$\hat{d}_i$	Distance measurement based on ToF at i-th step
$\tilde{d}_i$	Distance estimation at i-th step
σ_a	Anchor location estimation variance
σ_d	Distance estimation variance
φ_i	Orientation of the target and the anchor at i-th step
p	Target localization probability function
θ	The deterministic parameter of the localization function
z	Joint observation vector of the localization function
$f(\cdot)$	Objective function of the localization optimization

3.1.1 System Overview

iLoc is mainly composed of two components: an anchor node and target IoT nodes. As illustrated in Fig. 3.1, the anchor node contains a simplified gateway and a co-located commercial smartphone. To locate a target, the anchor node moves around to a set of locations during which it measures the distance and anchor locations for localization. Before presenting the detailed localization model, some important notations are summarized in Table 3.1.

3.1.2 Localization Model

For most wide-area IoT applications, we consider the localization in a two-dimensional space as shown in Fig. 3.2. In such a scenario, the target IoT node

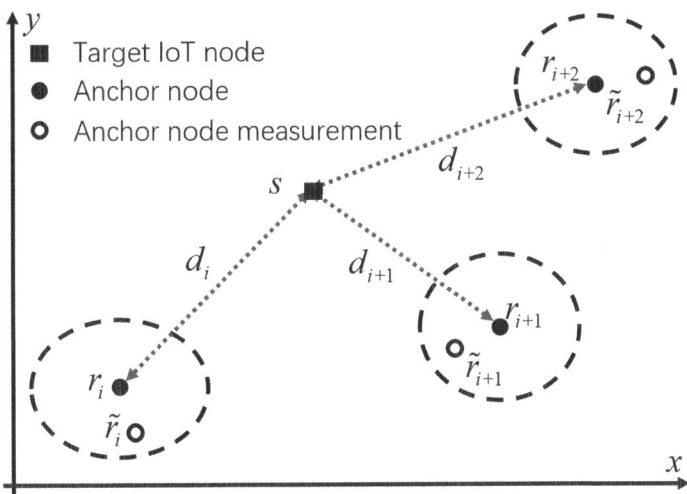

Fig. 3.2 Localization model

location is denoted by $s = [x, y]$, consisting of its x-axis and y-axis coordinates that can be transformed to the longitude and latitude in the map. The anchor node moves around and receivers its location measurement from the smartphone at each step, which is denoted by $\tilde{r}_i = [\tilde{x}_i, \tilde{y}_i](i = 1, 2, \ldots n)$. The measurement can be modeled as $\tilde{r}_i = r_i + e_i$, where r_i is the groundtruth. The noise e generated from the hardware and environment uncertainty is assumed to be Gaussian with zero mean and can be modeled as $e_i \in \mathcal{N}(\mathbf{0}, \mathbf{I}\sigma_a^2)$. The distance d_i between the anchor node and the target can be expressed as $d_i = \|s - r_i\|_2$. Hence, the orientation of the target, which is subtended at the anchor location is modeled as

$$\varphi_i = \tan^{-1}(\frac{y - y_i}{x - x_i}). \tag{3.1}$$

The estimated distance from the gateway is expressed as $\tilde{d}_i = d_i + o_i$, where o_i is the noise induced by the hardware and environment uncertainty. Without loss of generality, we assume that it has zero mean and is Gaussian, i.e. $o_i \in \mathcal{N}(0, \sigma_d^2)$ [1]. Once the distance measurements are observed in the moving path, they are expressed as a vector $\tilde{d} = [\tilde{d}_1, \tilde{d}_2, \ldots, \tilde{d}_n]$, and the groundtruth can be expressed as $d = [d_1, d_2, \ldots, d_n]$.

Hereby, when the anchor records an array of locations at different locations, the observation vector of locations can be denoted as $\tilde{r} = [\tilde{r}_1^T, \tilde{r}_2^T, \ldots, \tilde{r}_n^T]^T$. Correspondingly, the anchor node's location vector can be denoted by $r = [r_1^T, r_2^T, \ldots, r_n^T]^T$. Therefore, we can obtain the joint observation vector $z = [\tilde{d}, \tilde{r}]^T$, and the deterministic parameter $\theta = [s, r]^T$. The target IoT node location can be calculated by the following maximum likelihood estimation.

$$\tilde{\theta} = \arg\max_{\theta} = p(z|\theta) \tag{3.2}$$

3.2 Distance and Location Estimation

As the target IoT node location is estimated from the measurements including the distance and anchor location, we focus on the estimation and optimization of the distance and anchor location in this section.

3.2.1 Distance Estimation

When the LoRa gateway measures the distance from the target, it first sends a ranging request and then keeps timing until the response arrives. The distance is further calculated linearly via the timing result. However, due to the oscillator drift that leads the carrier frequency offset, the ranging results exhibit some non-linearity in the distance estimation at short distances, which degrades the ranging accuracy [2].

As a result, the distance estimation based on the LoRa ranging ToF is unstable in the short range, which makes iLoc unable to locate the target accurately when the user comes near to it. To solve this problem, we fuse Received Signal Strength Indication (RSSI) with ToF to mitigate the hardware uncertainty. The RSSI attenuates with the distance increase and is especially stable at short distances. In iLoc, when the gateway receives the responding packet, it can record the timing results and the RSSI as well. Hence, we fuse the polynomial regression model of ToF and the logarithmic distance path-loss model of RSSI to estimate the distance as the follow

$$\tilde{d} = \gamma \sum_{i=0}^{m} a_i \hat{d}^i + (1 - \gamma)10^{\left(\frac{R_0 - \hat{R}}{-10n}\right)}, \tag{3.3}$$

where γ, a_i, and m are regression parameters. n is the path-loss parameter, R_0 is the average measured RSSI at the distance of 1 m. $\hat{d}$ is the distance calculated from the ranging engine and $\hat{R}$ is the measured RSSI.

We collect the ranging results and RSSI measurements from 10 to 50 m to estimate the model parameters at short distances. The Mean Squared Error (MSE) is evaluated for different values of γ and m. We calculate the result and illustrate it in Fig. 3.3. Compared with the raw ToF measurements, the distance estimation becomes stable at short ranges and the median error is less than 1.3 m while the calculation based on ToF contains non-negligible errors. When the distance increases, the RSSI measurements become insensitive, but the ToF performance exhibits stability in such a scenario. As the ranging estimation performance is improved at short distances and exhibits stability at long distances, iLoc adopts a threshold to select the distance estimation model. If the raw ToF measurement is less than the threshold, it will be calibrated in the regression model to improve the distance estimation performance.

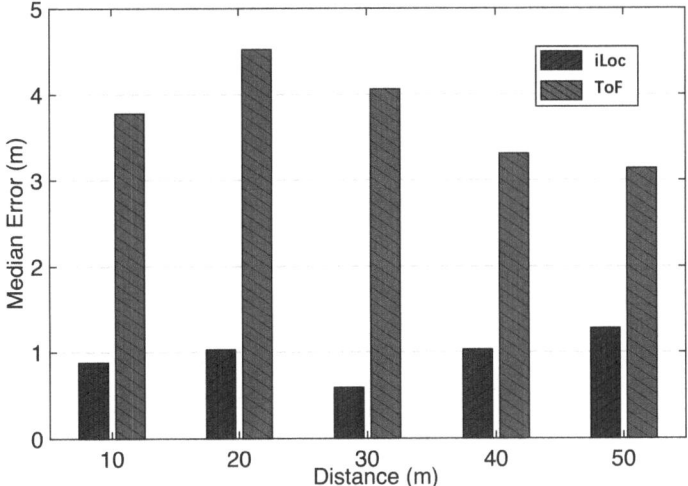

Fig. 3.3 Distance estimation

3.2.2 Anchor Location Estimation

Once the gateway obtains the distance estimation $\tilde{d}$, the anchor node records its location that is being estimated in the smartphone during the moving. Further, the measurements are fused in the target node localization. Hence, to locate the target with good performance, we need to locate the anchor node in the moving accurately. Fortunately, the commercial smartphones have been integrated with diverse sensors such as GPS receiver, IMU, and magnetometer for localization and navigation, which makes it possible for us to locate the anchor node during the moving without any extra hardware.

The existing smartphone localization system based on GPS receivers has been exploited in many applications [3, 4]. However, their localization resolution is coarse-grained and becomes unstable in building areas where some satellites are obstructed. In most urban IoT applications, the target localization is easily degraded by complex environments.

To mitigate the environmental uncertainty and GPS instability, we fuse GPS, IMU, and the magnetometer in the estimation, which are available in most commercial smartphones. The IMU module including an accelerometer and a gyroscope can measure the real-time attitude of the smartphone. Thus, the moving orientation can be estimated from the magnetometer and smartphone attitude no matter where the smartphone is held. Further, we estimate the path distance via the IMU module. In motion tracking systems, the Pedestrian Dead Reckoning (PDR) is a robust algorithm for IMU-based tracking and has been widely adopted in GPS-less systems [5, 6]. The location is calculated periodically based on the user steps. Therefore, the integral error in the IMU module can be mitigated at each step. Specifically, the

Weinberg method is an accurate and lightweight PDR model for mobile device, which calculates the step distance l from the difference between the maximum acceleration A_{max} and the minimum acceleration A_{min} [7].

$$l = k\sqrt[4]{A_{max} - A_{min}} \tag{3.4}$$

In the Weinberg method, k is the parameter that is correlated to the user's height and the place where the sensor is held. In iLoc, it is calculated with the GPS receiver in the pre-prepared calibration for users.

When the step distance and orientation are estimated, we can jointly estimate the user next step location, $\boldsymbol{p}_{k+1} = [x_{k+1}, y_{k+1}, l_{k+1}, \beta_{k+1}]^{\mathrm{T}}$ based on the current location $\boldsymbol{p}_k = [x_k, y_k, l_k, \beta_k]^{\mathrm{T}}$. To calculate the global location, the step measurements are fused with the GPS measurements via a particle filter. In this filter, the location is determined by step distance and orientation, which is further updated from the GPS receiver. We model the location estimation as

$$\boldsymbol{p}_{k+1} = \boldsymbol{p}_k + \begin{bmatrix} l_k \cos \beta_k \\ l_k \sin \beta_k \\ 0 \\ 0 \end{bmatrix} + \boldsymbol{w}_k$$

$$\boldsymbol{q}_k = \boldsymbol{C}\boldsymbol{p}_k + \boldsymbol{v}_k, \tag{3.5}$$

where β_k is the step orientation, $q_k = [longitude_k, latitude_k, l_k, \beta_k]^{\mathrm{T}}$, and $\boldsymbol{C}$ is the function that transforms the coordinate value to GPS data, w_k is the noise in the moving path, and v_k is the noise in GPS. The starting location $\boldsymbol{p}_0$ is set as initial GPS observation and then updated by 100 particles in each period. We select new particles at the range of 10–30 m near the estimation location, which is determined by the GPS accuracy.

We evaluate the anchor location estimation performance as illustrated in Fig. 3.4. The accuracy is significantly improved to sub-meters (median error 0.79 m) via the fusing of the IMU module, the magnetometer, and the GPS receiver. Based on the improved distance and anchor location estimation, we continue to optimize the moving path for the target IoT node localization, which will be introduced in the next section.

3.3 Localization Optimization

As iLoc localizes the target via the likelihood estimation, the localization performance is correlated to the virtual anchors' locations in the path. Without loss of generality, we minimize the estimation Cramer-Rao Lower Bound (CRLB) via scheduling the moving path and anchor locations to improve the accuracy. The CRLB inequality is normally characterized by the inverse of Fisher Information Matrix (FIM) that can be calculated by

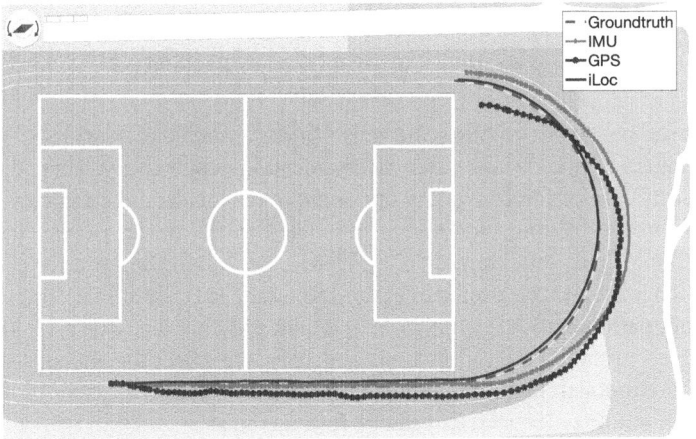

Fig. 3.4 Anchor location estimation

$$\mathbf{F}(s) = \mathbf{F}(s, s) - \mathbf{F}(s, r)\mathbf{F}(r, r)^{-1}\mathbf{F}^{\mathrm{T}}(s, r)$$

$$= \sum_{i=1}^{n} g_i \begin{bmatrix} \cos^2(\varphi_i) & \cos(\varphi_i)\sin(\varphi_i) \\ \cos(\varphi_i) & \sin^2(\varphi_i) \end{bmatrix}, \tag{3.6}$$

where g_i is the noise factor and can be expressed as $g_i = \frac{1}{\sigma_d^2 + \sigma_a^2}$. σ_a is the standard deviation of the anchor node location estimation . σ_d is the distance estimation standard deviation that can be derived from the standard deviation of the ToF σ_{tof}. In wireless ranging systems, the σ_{tof} is modeled as

$$\sigma_{tof}^2 = \frac{1}{8\pi^2 \cdot SNR \cdot \sqrt{K} \cdot BW^2 \cdot 2^{SF}}. \tag{3.7}$$

The SNR is the signal-to-noise ratio and the K is the number of ranging symbols [8]. Note that the ranging performance is correlated to its signal quality, which is sensitive to the communication distance. Generally, we calculate the SNR from the $RSSI$ that can be collected from the gateway. Further, the determinant of $\mathbf{F}$ can be calculated as

$$det\{\mathbf{F}\} = \frac{1}{2}\sum_{i=1}^{n}\sum_{i=1}^{n} g_i g_j \sin^2(\varphi_{ij}), \tag{3.8}$$

where $\varphi_{ij} = \varphi_j - \varphi_i$. Geometrically, the estimation error ellipse is $\pi\sqrt{\lambda_1 \lambda_2}$, and λ_1, λ_2 are eigenvalues of $\mathbf{F}^{-1}$. For simplicity, we drop the constant term π and express the estimation error bound e as

$$e = \sqrt{\frac{1}{det\{\mathbf{F}\}}} = \sqrt{\frac{2}{\sum_{i=1}^{n} \sum_{j=1}^{n} g_i g_j \sin^2(\varphi_{ij})}}. \tag{3.9}$$

The error bound e combines the noise factor g and the orientation φ, which are determined by the distance and anchor location, respectively. Thus, the virtual anchor location selection is not only important to the localization accuracy, but also significant to the moving path that is correlated to the power (longer path takes more working time of the target node). Hence, we minimize the error bound and the path length while the user moving to the target IoT node. The error bound e, the moving path length $\sum_{i=1}^{n} ||r_{i+1} - r_i||$, and the final distance d_n to the target are weighted in the objective function. Specially, we select the anchor location in practical environment that may contain unavailable obstacles areas in the map. The localization optimization problem can be formulated as

$$\min_{r} \quad f(r) = \lambda_1 e + \lambda_2 \sum_{i=1}^{n} ||r_{i+1} - r_i||$$

$$+ (1 - \lambda_1 - \lambda_2)d_n \tag{3.10}$$

$$\text{s.t.} \quad r_i \in M \quad i = 1, 2, 3 \ldots, n,$$

where M is the available areas of the map. λ_1 and λ_2 are the model parameters that will be calculated in the simulation. As the $f(r)$ is not a concave function, this problem is not convex. Thus, it is difficult to calculate the analytical solution. In iLoc, we focus on the next anchor location and design a heuristic path optimization method to obtain a minimum solution at each step. We just need to calculate the $t + 1$-th location at the current t-th location and iteratively follow the instruction until we find the target. Particularly, the $t + 1$-th objective function can be expressed as

$$f_{t+1}(r) = \lambda_1(m_t + g_{t+1} \sum_{i=1}^{t} g_i \sin^2(\varphi_{t+1} - \varphi_i))^{-\frac{1}{2}}$$

$$+ \lambda_2(d_t^2 + d_{t+1}^2 - 2d_t d_{t+1} \cos(\varphi_{t+1} - \varphi_t))^{\frac{1}{2}}$$

$$+ (1 - \lambda_1 - \lambda_2)d_{t+1}, \tag{3.11}$$

where m_t, d_t, g_i, and φ_i are the parameters that have been calculated in the last step. To minimize the $t + 1$-th object function, we adopt the Stochastic Gradient Descent (SGD) optimization to find a solution for the $t+1$-th location. The gradient of $f_{t+1}(r)$ on $r_{t+1} = [x_{t+1}, y_{t+1}]$ is complex while we can calculate it on the angle φ_{t+1} and distance d_{t+1} simply. At the current location, we set r_t as the initial value and update the calculation as

$$\begin{bmatrix} \varphi_{i+1} \\ d_{i+1} \end{bmatrix} = \begin{bmatrix} \varphi_i \\ d_i \end{bmatrix} - \mu(\nabla f_{t+1}([\varphi_i, d_i]^{\mathrm{T}}) + \epsilon_i), \tag{3.12}$$

where ∇f_{t+1} is the gradient function, ϵ is the stochastic gradient noise, and μ is the step factor. Hence, we first calculate the distance and the angle via the iterative method and then infer the next location r_{t+1} as the follow

$$r_{t+1} = r_t + L_{t+1} \begin{bmatrix} \cos(\alpha_{t+1}) \\ \sin(-\alpha_{t+1}) \end{bmatrix}, \qquad (3.13)$$

where L_{t+1} and α_{t+1} are the geometric parameters that demonstrate the next location and can be derived from the gradient optimization results. When the next step location falls into the obstacle area during the iteration, iLoc stops the iteration and returns the minimum value in the record.

We design the localization optimization algorithm as illustrated in Algorithm 1. iLoc selects the first and the second steps randomly based on the initial anchor location and leads the user to the locations. At the same time, it estimates the anchor

---

Algorithm 1: Localization optimization

Input: maximum times of calculation $iter_{max}$ and t_{max}, the first and second locations r_2, r_3 that are generated randomly to travel to, iterative threshold h, the step factor μ, and the model weights λ_1, λ_2, the distance threshold d.

Output: the target IoT node location $\tilde{s}$

Collect distance $\tilde{d}_1$ at initial anchor location $\tilde{r}_1$;

Generate the first and second locations $\tilde{r}_2$, $\tilde{r}_3$;

Collect $\tilde{d}_2$ and $\tilde{d}_3$ at $\tilde{r}_2$ and $\tilde{r}_3$, respectively;

Initialize iterative error $e_{iter} = h$ and minimal records $f_{min} = f_3([\varphi_3, d_3])$, $\varphi_{min} = \varphi_3$, $d_{min} = d_3$, $t = 3$;

while $\hat{d}_t < d$ *or* $t \geq t_{max}$ **do**

    Initialize $iter = 0$, $d_{t+1} = \tilde{d}_t$, $\varphi_{t+1} = \tilde{\varphi}_t$;

    while $e_{iter} \geq h$ *and* $iter \leq iter_{max}$ **do**

        Update d_{t+1} and φ_{t+1} via (3.12);

        Calculate f_{t+1} via (3.11);

        if $f_{min} \leq f_{t+1}$ **then**

            $f_{min} = f_{t+1}$, $\varphi_{min} = \varphi_{t+1}$, $d_{min} = d_{t+1}$;

        Calculate r_{t+1} via (3.13);

        if $r_{t+1} \notin \mathbf{M}$ *or* $iter > iter_{max}$ **then**

            $\varphi_{t+1} = \varphi_{min}$, $d_{t+1} = d_{min}$;

            Update r_{t+1} via (3.13);

            break;

        Calculate iterative error e_{iter} from the difference of f_{t+1} in two times iteration;

        $iter = iter + 1$;

    end

    Estimate the anchor location and move to r_{t+1};

    Estimate the distance $\tilde{d}_{t+1}$;

    Update the target location estimation $\tilde{s}$ via (3.2);

    $t = t + 1$;

end

return $\tilde{s}$;

---

node location and the distance from the target node. Then, it starts optimizing the path for the next step. To calculate the minimal solution, the current estimated $\tilde{\varphi}_t$ and $\tilde{d}_t$ are calculated as the initial value. Further, those variables are iteratively updated according to the gradient, during which the minimal f_{min} is recorded. If the next anchor location falls outside of the accessible map, the minimal φ_{min} and d_{min} are returned as the next step location. When the next location is obtained, iLoc continues to lead the user to the optimized locations and keeps estimating the anchor node location. As a better virtual anchor will be established in r_{t+1}, the target IoT node location estimation is updated with $\tilde{r}_{t+1}$ and $\tilde{d}_{t+1}$. Hence, the localization accuracy is heuristically improved during the moving. The target location is returned when the estimated distance $\tilde{d}_t$ is less than the threshold, which means the user has been guided to find the target in the map.

3.4 Implementation and Evaluation

In this section, we present the system implementation and performance evaluation. Firstly, we introduce the system design including the LoRa tag, the anchor node, and the localization framework. Then we conduct simulations and experiments in different environments to evaluate the system performance.

3.4.1 System Implementation

iLoc is mainly composed of two components: the target IoT node and the mobile anchor that includes a simplified gateway and a commercial smartphone. The target IoT node is designed as a lightweight tag that can be attached to the machines, packages, and architectures in IoT applications. The gateway connects to the co-located smartphone via Bluetooth Low Energy (BLE) and transmits the measured distance to the localization module in the localization framework. We developed the framework for localization and navigation on the Android platform and evaluate it on commercial smartphones.

3.4.1.1 LoRa Tag Design

LoRa is a popular LPWAN platform that has been adopted in many IoT applications [16, 17]. However, there are also some devices that are not equipped with LoRa modules. For those nodes and new emerging IoT applications, we design a lightweight LoRa tag that only consists of basic components for communication and localization. The microcontroller STM32L011 (0.9\$), LoRa chip SX1280 (2.9\$), and the printed antenna are integrated on a tiny printed circuit board (16 mm $\times$ 33 mm). Thus, the tag is small enough to be attached to target devices in IoT applications. The simplified tag costs less than 5 dollars and is affordable for most IoT applications. We design the ranging procedures to comply with the standard frequency hopping protocol to simplify the tag deployment. When receiving a ranging request packet, the tag decodes the address of the gateway and selects a channel frequency randomly for the next ranging hopping. Then, its address and the selected channel are encoded in the response packet that is transmitted to the gateway.

3.4.1.2 LoRa Gateway Design

The LoRa gateway is a core component in the anchor node, which estimates the distance and connects the target node with the IoT cloud platform. The gateway is composed of a LoRa chip, a microcontroller STM32L476, a BLE module CC2640, and a lithium battery (600 mAh). We remove the expensive ethernet network module, compute module, and display module, which decreases the manufacturing cost to less than 10 dollars. The simplified structure is also lightweight for mobile applications. When the user selects the target node in the smartphone, its address is transmitted to the gateway. Then, the gateway sends a ranging request packet. When the LoRa gateway receives the response packet from the target, the microcontroller decodes the messages and synchronizes the hopping frequency for the next communication. Meanwhile, the collected ToF and RSSI are fused in the distance estimation, which is transmitted to the smartphone via the BLE module.

3.4.1.3 Localization Framework Design

The localization framework connects the LoRa gateway and the embedded sensors as illustrated in Fig. 3.5. The framework is mainly composed of an anchor location estimation module, a target localization module, and a communication module. The communication module connects to the gateway to collect the node's data and transmits them to the IoT cloud platform. The distance estimation module calculates the distance from the target when receiving ToF and RSSI measurements from the communication module. Meanwhile, it sends a signal to the anchor location estimation module that calculates the location via fusing the GPS receiver, the IMU sensor, and the magnetometer. Further, the distance and the anchor location are sent to the target localization module that demonstrates the results and navigation to the user in the smartphone.

3.4.2 Localization Simulations

We run the simulations to verify the efficiency of our iterative localization algorithm. Firstly, we consider an open 2 km× 2 km area with the randomly deployed target

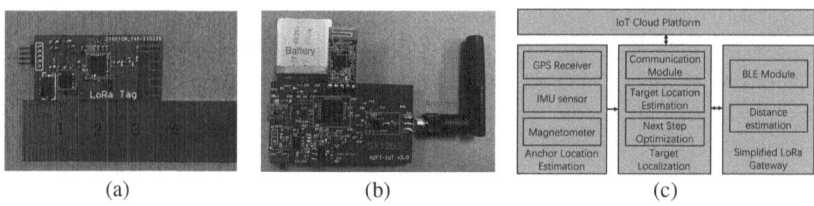

Fig. 3.5 Hardware and framework design. (**a**) LoRa tag. (**b**) Simplified mobile LoRa gateway. (**c**) Localization framework

IoT node. Without loss of generality, an anchor node starts from the original location in the area. The distance measurements are emulated from the results in Sect. 3.2.1 and the anchor location estimation error is calculated from the results in Sect. 3.2.2. We set the step factor $\mu = 0.05$, the path length factor $\lambda_2 = 0.05$, and the error bound factor $\lambda_1 = 0.3$, 0.35, and 0.4 in three simulations. The anchor node follows the optimization results iteratively. It stops moving if the collected distance is less than 5 m where we can find the target device generally.

3.4.2.1 Simulation Results

We test the optimized path that saves the power consumption and improves the localization accuracy. To compare the paths simulated on different parameters, the target is placed in the center of the area. We run tests for the three parameters mentioned above. The optimization starts from the third step while the first step, $(50, 0)$, and the second step, $(50, 50)$, are selected as the same to avoid the initialization interference. The paths are illustrated in Fig. 3.6. The anchors move to the target iteratively and all locate the target finally. The anchors take 19 steps in the moving to locate the target with the error standard deviation of 5.00 m, 4.51 m, and 4.26 m in Fig. 3.6a, b, and c, respectively. The simulation comparison demonstrates that the parameter λ_1 determines the direction biased from the target. This is mainly because the anchor tries to collect more widely distributed samples when the target location is uncertain at the beginning. However, the wide direction leads to long path length while it improves the accuracy. Our algorithm keeps the balance between localization accuracy and path length, which is important in IoT applications.

To further verify the optimization performance, we resize the test area to 400 m $\times$ 400 m. The target is also placed at the central, and the anchor starts from the origin. The first and the second locations are selected at $(10, 0)$ and $(10, 10)$. We can find similar features of the paths in Fig. 3.7 compared with Fig. 3.6. The anchors locate the target via 16 steps in the path with the error standard deviation of 2.20 m, 1.68 m, and 1.28 m in Fig. 3.7a, b, and c, respectively. The different area simulations demonstrate iLoc can adaptively select the steps in different ranges.

We also test the localization error bound and the path length in the simulations, which is correlated to the power consumption. We test two methods in localization.

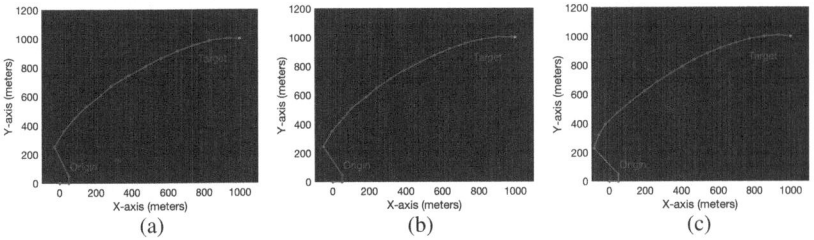

Fig. 3.6 Optimized path in 2 km area. (**a**) $\lambda_1 = 0.30$. (**b**) $\lambda_1 = 0.35$. (**c**) $\lambda_1 = 0.40$

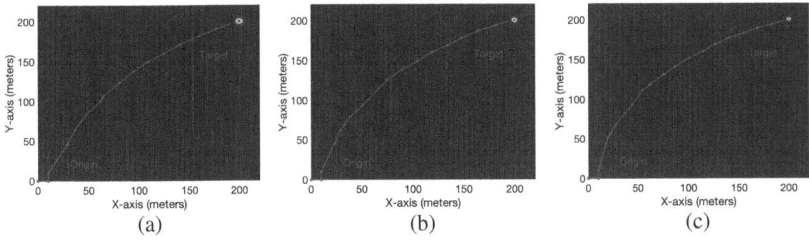

Fig. 3.7 Optimized path in 400 m area. (**a**) $\lambda_1 = 0.30$. (**b**) $\lambda_1 = 0.35$. (**c**) $\lambda_1 = 0.40$

The first is random step selection, and the second is iLoc. The step length sum and the error bound are recorded at each step. The results are compared in a 400 m × 400 m area. The parameter λ_1 that determines the error bound is 0.4. As presented in Fig. 3.8, the error bound decreases during the anchor moving. This is because the anchor collects more ranging samples for localization. In the random selection method, the error bound decreases to less than 20 m when the anchor moves 250 m long. However, iLoc achieves such a level with just 100 m. This demonstrates that iLoc's efficiency is 2.5 times of the random method. In the optimized path, the error bound becomes less than 0.7 m at 300 m while it's still over 10 m in the random method.

The simulations illustrate the efficiency and accuracy of our system. To verify its performance in real scenarios, we conduct field experiments to evaluate localization accuracy and power consumption.

3.4.3 System Evaluation

In this subsection, we first investigate the popular system's cost and then carry out numerous field experiments with a LoRa gateway, a LoRa tag, and a commercial smartphone.

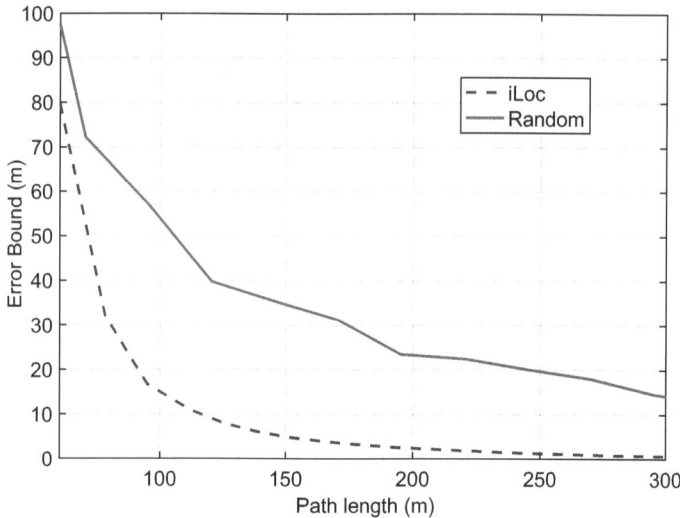

Fig. 3.8 Error bound vs. path length

Table 3.2 Manufactory cost of four systems

Techniques	GPS	WiFi	SLAM	iLoc
Sensor	GPS receiver	WiFi transceiver	Camera	LoRa transceiver
Sensor cost ($)	20–30	5–20	30–200	3–5
Effective range	Wide area	10–50 m	Wide area	Wide area
Localization error (m)	≤ 10	≤ 5	≤ 20	≤ 3
Infrastructure support	Satellites	Access point	None	Anchor (10$)

3.4.3.1 System Cost

We investigate the hardware manufacturing cost and performance of several popular localization technologies. In the comparison, only the node is discussed as illustrated in Table 3.2 since many anchors have been deployed in the environments (e.g. satellites, base stations).

Localization technologies based on GPS and Simultaneous Localization and Mapping (SLAM) incur much more manufacturing costs than WiFi and iLoc. The cost of the LoRa chip can be as low as 3 dollars while the effective range and localization error are also acceptable. Its low cost is significant in numerous wide-area IoT applications. Moreover, iLoc can be directly applied to the IoT applications that have been equipped LoRa modules. In such a scenario, iLoc incurs no extra overhead in the deployments.

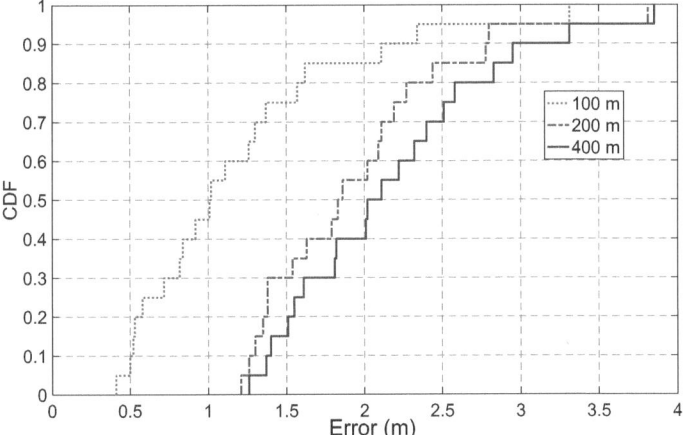

Fig. 3.9 CDF of localization error in different ranges

3.4.3.2 Localization Performance

We conduct numeral experiments with the implemented prototype and a smartphone in different areas to evaluate the localization accuracy.

3.4.3.3 Testbed

In the experiments, the LoRa tag is randomly placed in areas of 100, 200, and 400 m. The user holds the smartphone, Samsung Galaxy A9 Star, and the simplified gateway to follow the instructions from the algorithm that calculates the next steps to collect the ranging samples. The smartphone is integrated with a general Global Navigation Satellite System (GNSS) location hub chip, a magnetometer, and an IMU sensor, which support the anchor location estimation. We run the gateway in LoRa ranging master mode while the target is in slave mode. We select the 1625 kHz bandwidth and the spreading factors is set as SF6. The transmitting power is 18 mW.

3.4.3.4 Results

The user tests the localization experiment 20 times in each area. The localization error is recorded at each experiment and the Cumulative Distribution Function (CDF) is illustrated in Fig. 3.9.

In the range of 100 m, iLoc achieves the mean localization error of 1.19 m with a variance of 0.72 m. Above all measurements, 80% errors are less than 1.6 m. In the area of 200 m, the mean localization error is 1.95 m, and 90% errors are less than 2.8 m. Similarly, in the area of 400 m, the mean localization error is 2.17 m,

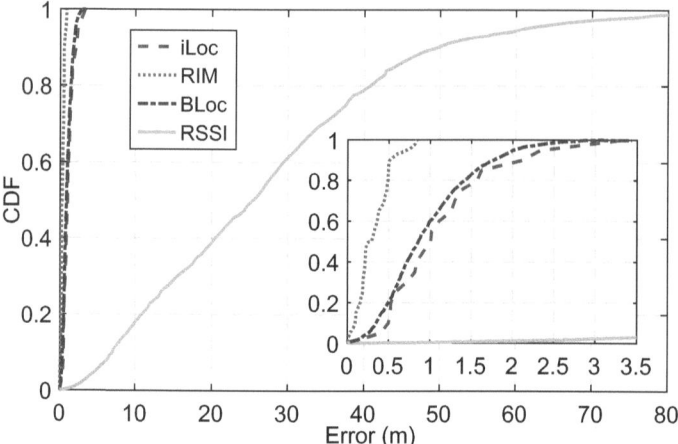

Fig. 3.10 CDF of localization error comparison

and 90% errors are less than 3.3 m. iLoc performs better in an area of 100 m due to the improved distance estimation. In 200 and 400 m, the results are similar because the distance estimation in a wide area is based on the ToF that is stable in the long range, and the optimization algorithm can adaptively select the steps for the user in different ranges.

We further compare the performance with some state-of-the-art systems including the WiFi-based RIM, the Bluetooth-based BLoc, and the LoRa-based RSSI system [9–11].

As presented in Fig. 3.10, the WiFi-based RIM achieves an accurate performance with 0.3 m median error via a single virtual antenna. The proposed localization algorithm improves the location resolution significantly based on the MIMO WiFi device. For the nodes with one antenna, BLoc collects the Channel State Information (CSI) in the anchor and achieves a localization accuracy of 0.86 m. While the CSI is unavailable in LoRa, iLoc fuses multiple sensors and optimizes the virtual anchor locations to improve localization accuracy. The median error is 1.19 m and is close to BLoc. Compared with the LoRa RSSI localization system, it improves the accuracy 10 times. Though its accuracy is not as high as WiFi-based techniques, meter-level accuracy is acceptable by many wide-area IoT applications.

3.4.3.5   System Robustness Evaluation

To evaluate the performance of our system in various scenarios for different IoT applications, we conduct experiments in three representative environments: namely playground, bush, and building. The user starts from the selected origin location and follows the optimized steps to locate the target for 20 times experiments. The CDF of localization errors for three environments is presented in Fig. 3.11

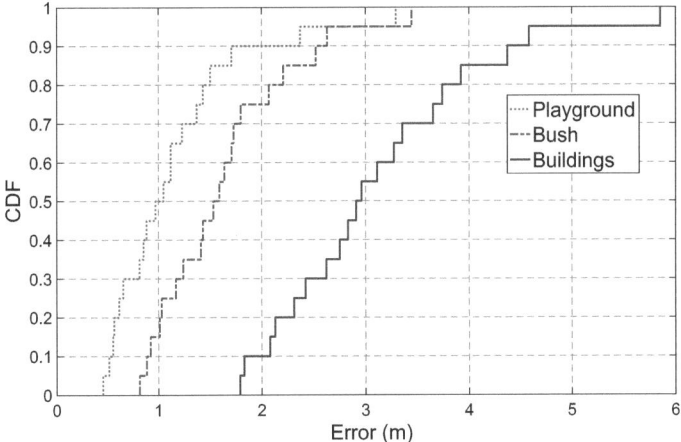

Fig. 3.11 CDF of localization error in different scenarios

In the playground, we can see that 90% localization errors are less than 2.5 m. Since the playground is an open environment without obstruction, the ranging distance can be measured more accurately. The localization accuracy in bush environments has a similar performance as those in the playground. The average error is 1.64 m and 90% errors are less than 3 m. This is mainly because the obstruction in the bush environment is small and the judiciously selected locations for the anchor node by iLoc can alleviate its impact. We also evaluate system performance in an area with some tall buildings. In this case, the target is always obscured by the buildings that lead to a non-line of sight distance measurement. Moreover, GPS data is sometimes inaccessible, which degrades the anchor location estimation. iLoc overcomes those obstacles via the fused sensors and the optimized path in the localization. It achieves an average localization error of 3.12 m while 90% localization errors are still acceptable for wide-area IoT applications.

3.4.3.6 Power Consumption Evaluation

The localization systems designed for IoT applications are sensitive to power consumption since most IoT devices are equipped with a limited battery. In this subsection, we conduct experiments to evaluate the power consumption of iLoc and compare it with the existing localization techniques. We focus on the target node energy consumption since it is costly to recharge them once they are deployed in IoT applications. In the experiment, we record the power for 5 minutes and calculate the average consumption via a power monitor. The comparison results are illustrated in Table 3.3.

As presented in Table 3.3, iLoc is compared with the Wi-PoS (UWB localization), visible light localization system, and LoRa-based localization system. iLoc

Table 3.3 Power consumption comparison

Systems	Technology	Power
WiFi NIC [12]	Wi-Fi: RSSI	500.0 mW
Wi-PoS [12]	UWB: ToF	189.0 mW
GPS receiver [13]	GPS: TDoA	177.0 mW
ZiLoc [14]	ZigBee: RSSI	167.0 mW
Pulsar [15]	Visible light: AoA	150.0 mW
GPS-less [13]	LoRa: RSSI	80.0 mW
iLoc	LoRa: ToF&RSSI	53.5 mW

takes a power consumption of 53.5 mW that is only 28% of the Ultra Wide Band (UWB) localization system (it has a very high localization accuracy with a short coverage range) and 10.7% of the Wi-Fi localization system. Particularly, iLoc even takes less power consumption compared with the WLAN technologies [14]. Because iLoc only collects samples at the selected steps. The proposed algorithm decreases the path length and saves the localization time. We consider 5 minutes walking for the localization generally, such that the power consumption is 1.06 mAh, which is ignorable for most IoT nodes.

3.5 Summary

In this paper, we have proposed iLoc, a novel low-cost, low-power, and wide-area mobile localization system for IoT applications. Instead of relying on deployed LoRa stations, we designed a simplified mobile LoRa gateway, which cooperates with a commercial smartphone to imitate an array of anchors in the localization. The LoRa gateway costs less than 10 dollars and the LoRa tag costs less than 5 dollars. To improve localization accuracy, we fuse the ToF and RSSI measurements in the distance estimation and exploit multiple sensors in the anchor location estimation. Particularly, we design an optimization algorithm that improves the localization accuracy and reduces the power consumption as well. The proposed system chooses a set of locations heuristically and illustrates the path to find the target for users. Numerous simulations and field experiments are conducted, which demonstrate that iLoc has a good localization performance for wide-area IoT applications.

References

1. M. Jadaliha, Y. Xu, J. Choi, N.S. Johnson, W. Li, Gaussian process regression for sensor networks under localization uncertainty. IEEE Trans. Signal Process. **61**(2), 223–237 (2012)
2. Semtech. SX1280, Accessed: 2022 [Online]. Available: https://www.semtech.com/products/wireless-rf/24-ghz-transceivers/sx1280
3. K. Chen, G. Tan, BikeGPS: localizing shared bikes in street canyons with low-level GPS cooperation. ACM Trans. Sens. Netw. **15**(4), 1–28 (2019)

4. O. Kaiwartya, Y. Cao, J. Lloret, S. Kumar, N. Aslam, R. Kharel, A.H. Abdullah, R.R. Shah, Geometry-based localization for GPS outage in Vehicular cyber physical Systems. IEEE Trans. Veh. Technol. **67**(5), 3800–3812 (2018)
5. Y. Duan, K.Y. Lam, V.C. Lee, W. Nie, H. Li, J.K.Y. Ng, Packet delivery ratio fingerprinting: toward device-invariant passive indoor localization. IEEE Internet Things J. **7**(4), 2877–2889 (2020)
6. Z. Li, X. Zhao, F. Hu, Z. Zhao, J.L.C. Villacrés, T. Braun, SoiCP: a seamless outdoor–indoor crowdsensing positioning system. IEEE Internet Things J. **6**(5), 8626–8644 (2019)
7. Analog devices. AN-602 Application note, Accessed: 2022 [Online]. Available: https://www.analog.com/media/en/technical-documentation/application-notes/513772624AN602.pdf
8. S. Lanzisera, D.T. Lin, K.S. Pister, RF time of flight ranging for wireless sensor network localization, in *Proceedings of IEEE WISES, Vienna*, Jun 2006
9. C. Wu, F. Zhang, Y. Fan, K.R. Liu, RF-based inertial measurement, in *Proceedings of ACM SIGCOMM, Beijing*, Aug 2019
10. R. Ayyalasomayajula, D. Vasisht, D. Bharadia, BLoc: CSI-based accurate localization for BLE tags, in *Proceedings of ACM CoNEXT, Heraklion*, Dec 2018
11. K.H. Lam, C.C. Cheung, W.C. Lee, RSSI-based LoRa localization systems for large-scale indoor and outdoor environments. IEEE Trans. Veh. Technol. **68**(12), 11778–11791 (2019)
12. B.V. Herbruggen, B. Jooris, J. Rossey, M. Ridolf, N. Macoir, Q.V. den Brande, S. Lemey, E. D. Poorter, Wi-PoS: a low-cost, open source ultra-wideband (UWB) hardware platform with long range sub-GHz backbone. Sensors **19**(7), 1548 (2019)
13. A. Mackey, P. Spachos, LoRa-based localization system for emergency services in GPS-less environments, in *Proceedings of IEEE INFOCOM WKSHPS, Paris* (2019)
14. J. Niu, B. Lu, L. Cheng, Y. Gu, L. Shu, ZiLoc: energy efficient WiFi fingerprint-based localization with low-power radio, in *Proceedings of IEEE WCNC, Paris*, Apr 2013
15. C. Zhang, X. Zhang, Pulsar: towards ubiquitous visible light localization, in *Proceedings of ACM MobiCom, Snowbird*, Oct 2017
16. H.C. Lee, K.H. Ke, Monitoring of large-area IoT sensors using a LoRa wireless mesh network system: design and evaluation. IEEE Trans. Instrum. Meas. **67**(9), 2177–2187 (2018)
17. L. Leonardi, F. Battaglia, L.L. Bello, RT-LoRa a medium access strategy to support real-time flows over LoRa-based networks for industrial IoT applications. IEEE Internet Things J. **6**(6), 10812–10823 (2019)

Chapter 4
Wide-Area Localization System Based on LoRa Mesh

Abstract The positioning task of the Internet of Things (IoT) for outdoor environment requires that the node devices meet the requirements of low power consumption, long endurance and low cost, and that the positioning system can achieve high-precision positioning and wide-area coverage. Traditional IoT positioning technology cannot balance the cost, energy consumption and positioning performance well. Based on LoRa technology, LoRaWAPS, a low-cost, low-power and outdoor positioning system with multi-anchor wireless Mesh networking and multi-dimensional data fusion is designed in this chapter. To meet the need of positioning system, a low-power consumption and high-reliability LoRa Mesh protocol is proposed. Aiming at the problem that the accuracy of LoRa ranging is easily affected by the non-line-of-sight path propagation of signals, a distance estimation algorithm based on the fusion of time of flight (ToF) and received signal strength indicator (RSSI) multi-sampling data is proposed. Furthermore, a position estimation algorithm is designed to minimize a posteriori RSSI error for multi-anchor cooperative estimation scenarios. Based on the LoRa Mesh protocol, and the ranging and positioning algorithm, the prototype of LoRaWAPS is built and tested in the campus environment. The experimental results show that the proposed system can provide good location service for campus area. The peak power consumption of a single device in the system is less than 120 mW and the cost is less than $10, which can meet the outdoor positioning requirements of low-power and low-cost.

Keywords Low-power wide-area Internet of Things networking · LoRa · Outdoor positioning · Mesh network · Time-of-flight ranging

This chapter introduces LoRaWAPS, a LoRa Mesh based wide-area positioning system (LoRaWAPS). LoRaWAPS aims to establish an extensive network of affordable and energy-efficient positioning anchors interconnected through a mesh topology, ensuring comprehensive signal coverage. The system architecture is illustrated in Fig. 4.1. Proximity anchors located near the target are responsible for measuring the distance to the target and transmitting the ranging results back to the gateway and server via the mesh network. Subsequently, the server calculates

Fig. 4.1 System architecture

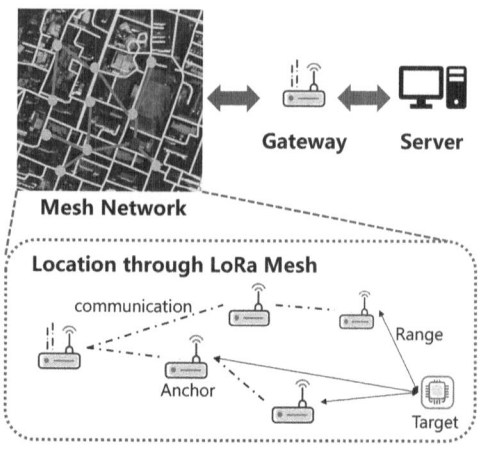

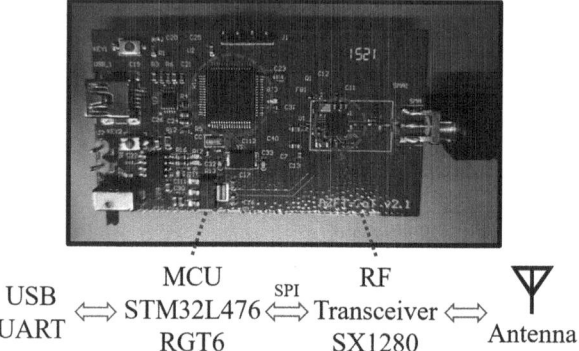

Fig. 4.2 The hardware structure of the target

the precise location of the target. In this way, LoRaWAPS reduces the blind area obscured by buildings and provides wide-area localization services.

In this chapter, we first present the design of the hardware and then illustrate the detail of the mesh networking. Then, the ranging and localization algorithms are proposed. Finally, we experimentally evaluate the proposed approach.

4.1   Hardware Design

The system consists of gateways, anchors, and targets. The gateway and anchor consist of a LoRa radio module, a control module, an interface module, a Low Noise Amplifier (LNA), and a high-gain antenna. The hardware structure of the target is similar to the anchor, except that the LNA is removed, as shown in Fig. 4.2.

The radio module uses the SX1280 chip, an RF integrated chip developed by Semtech, which can carry out ultra-long distance communication in the 2.4 GHz band. The receiving gain of LNA is 10 dB and the transmitting gain is 11 dB. It has a built-in automatic discrimination circuit for high-speed transceiver switching, which can automatically switch the transceiver state and work automatically without the control signal given by the control chip. The control module is based on STM32L476RGT6, a low-power ARM Cortex-M4 microprocessor. The control module is responsible for controlling the RF module and uploading the data. The control module is connected to the radio module through the serial peripheral interface (SPI). The interface module uses the serial port to the network port chip to provide wired network access and data reporting (as a gateway).

The current design focuses on functional verification, so its size and low-power characteristics are not optimized. Despite that, all following algorithms are device-independent by adding a hardware abstraction layer, making it easy to port to other platforms for cost control or low-power optimization.

4.2   LoRa Mesh Protocol Design

In this section, we describe the design of the mesh networking protocol. The Mesh networking protocol is an on-demand routing protocol. It discovers and maintains routes only when data needs to be sent, effectively reducing power consumption. We also designed the layered structure of the grid protocol to make it modular and maintainable. Furthermore, we propose a routing algorithm that balances efficiency and communication quality.

4.2.1   The Structure of the Mesh Protocol

The proposed Mesh protocol consists of four layers: physical layer, link layer, routing layer, and application layer. Its architecture is shown in Fig. 4.3.

Fig. 4.3 The workflow of the mesh protocol

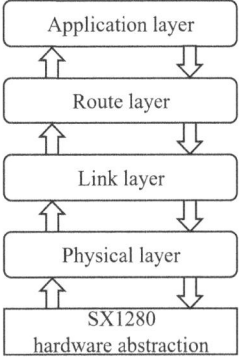

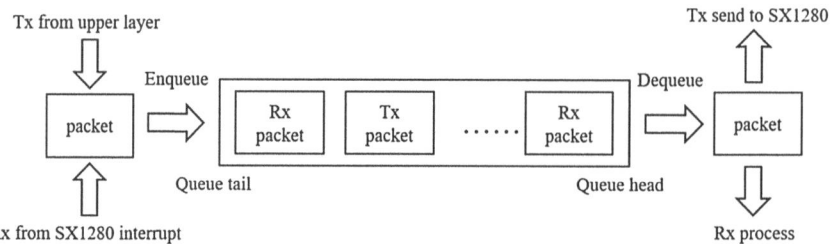

Fig. 4.4 The packet queue in physical layer

The physical layer of the system establishes communication with the hardware and ensures network reliability at the hardware level. To optimize packet handling and maintain packet order while minimizing hardware conflicts, a packet queue is implemented within the physical layer, as depicted in Fig. 4.4. This queue effectively manages the serialization of both transmitted and received packets. In order to mitigate data collisions, the physical layer employs two key mechanisms: Channel Activity Detection (CAD) and a random delay mechanism. Prior to transmitting data, devices are required to activate CAD mode to check the channel's occupancy. CAD utilizes LoRa lead codes to detect wireless channel activity. These lead codes, typically 12 symbols, assist the receiver in synchronizing with the incoming data stream. If the channel is found to be occupied, the device can either switch to a different channel or wait for a random period of time before attempting data transmission. Furthermore, the system employs a 32-bit Cyclic Redundancy Check (CRC32) to enhance data integrity. The CRC32 algorithm helps identify and reduce data errors, ensuring the accuracy of transmitted information.

Link layer ensures that packets passing to upper layers are all packets to be processed by this node. In this layer, we defined the target and source IP in this specific hop, which only represents the previous hop node and the next hop node of the packet. The node will only process the packet whose target IP is the same as its own IP or is a broadcast IP. To prevent the infinite flooding of a broadcast packet, link layer also maintains a list of nodes traveled.

Route layer is the key layer of our mesh protocol, maintaining a route table, executing the routing algorithm, and ensuring the best effort delivery of the packet. We use a routing table that contains route information on the network collected from received packets. When a packet is to be sent, the route layer protocol will choose a path from the route table, and create a route layer header. We also maintained a waiting list in this layer, which is used to handle replies and retransmissions. If no reply is received in a specified resend interval time, a packet without a reply might be retransmitted several times before declaring an error.

The topmost layer is application layer, whose role is to implement functions such as sending positioning instructions, handling messages, or communicating with the host computer.

The packet structure is shown in Fig. 4.5, each layer will add a new header. In the physic layer header, LoRa ID is used to distinguish LoRa packets that belong to our

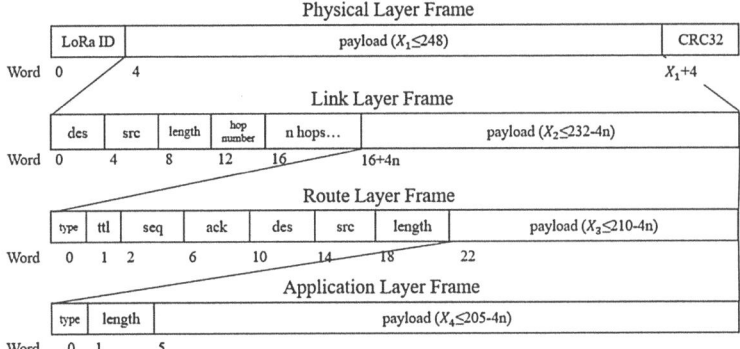

Fig. 4.5 The packet structure

system and CRC32 is used to check the payload's Integrity. In the link layer header, destination IP and source IP node IPs in this specific hop, length field represents the payload's length, and hop number records the number of nodes the packet has passed through. Following the hop number filed, a list records all nodes traveled. Route layer header has 7 fields, destination IP and source IP indicates the final and initial nodes which differ from those in link layer. Application layer packet only contains the type and length field for content indication.

The packet structure, as depicted in Fig. 4.5, incorporates headers added by each layer of the system. Starting with the physical layer header includes the LoRa ID, which serves to distinguish LoRa packets specific to our system, and the CRC32, which verifies the integrity of the payload. Moving to the link layer header, it contains the destination IP and source IP, representing the IP addresses of the nodes involved in the current hop. The length field indicates the length of the payload, while the hop number keeps track of the number of nodes the packet has traversed. Additionally, a list is maintained to record all the nodes the packet has passed through. In the route layer header, there are seven fields. The destination IP and source IP in this layer refer to the final and initial nodes, respectively, which may differ from those in the link layer. Finally, the application layer packet comprises two fields: the type field, which indicates the type of content carried by the packet, and the length field, which specifies the length of the content.

4.2.2   Route Discovery and Maintenance

As mentioned before, our LoRa mesh protocol is a kind of on-demand routing protocol, when there are no positioning tasks, all nodes keep in an idle state. When there is any job, the network starts the route update and maintenance process. As shown in Fig. 4.6, when start transmitting a message, the node will look up the route table to find a valid route to its destination, if the route exists, the packet will be

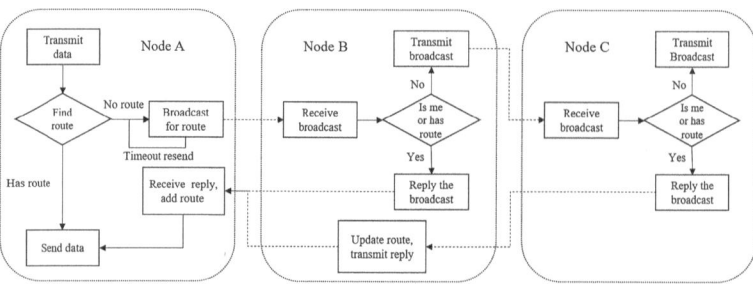

Fig. 4.6 Flowchart of route lookup

transmitted following the path, if not, the node will start finding a route by sending a broadcast lookup packet. Each node that received the broadcast packet but has no valid route to the destination will relay this broadcast. If received by the destination node or any node with a valid route to the destination, they will reply to the source node according to the path of the broadcast packet. After sending the lookup packet, the source node will add it to the waiting list for timeout re-transmission. Once the lookup packet reply is received, the suspended message packet will be transmitted to the destination through the discovered route.

When the network is active, each node can update its route table by capturing the packets of the surrounding nodes. Each route table entry has a valid time field, which will be refreshed by a new packet from the route path, and the RSSI field can also be updated by any captured packet. If the entry's valid time ends, the entry will be set to invalid. If one route entry failed on several transmission tasks, it will also be set to invalid.

4.2.3 Routing Algorithm

After receiving the broadcast packets forwarded by multiple nearby nodes, the node to be added to the network needs to select the most appropriate node to join the network. There are two main routing methods in current Mesh networking protocols. One method is to choose the node with the lowest signal attenuation between themselves, namely the method of maximizing RSSI. The other method is to select the node with the smallest hop count to the gateway, that is, to minimize the hop count. However, these two routing algorithms cannot balance communication efficiency and communication quality well. The goal of minimal hop count networking is to reduce the hop count of routes. However, when the RSSI between two nodes is lower than −80 dBm, the link reliability deteriorates seriously, which is reflected in a high packet loss rate. Maximizing RSSI Select the path with the largest overall RSSI in the link to ensure the connection quality between each hop. However, if the hop limit is not considered, the selected path may have unnecessary forwarding, which increases the transmission delay.

A heuristic algorithm is designed to find the optimal route in order to take into account node load, Packet Delivery Ratio, and Data Transmission Time (DTT). First, we modified LoRa broadcast packets. When each node forwards broadcast packets, the hop count of the node to the gateway and the current load of the node is updated. After receiving the broadcast packets forwarded by surrounding nodes, the node to be connected to the network can calculate the attractiveness of surrounding nodes. Attractiveness can be calculated based on self-measured RSSI and the information in received packets. We define the attraction of node i to node j as follows.

$$K_{i,j} = \frac{r_{i,j}^2}{M_j^{T_j}} \tag{4.1}$$

$$r_{i,j} = \begin{cases} 10^{\frac{-60}{10}}, & r_{i,j} > -60\,\text{dBm} \\ 10^{\frac{r_{i,j}}{10}}, & r_{i,j} \leq -60\,\text{dBm} \end{cases} \tag{4.2}$$

In the equations, $K_{i,j}$ is the attraction of node j to node i. Node i will choose a surrounding node j with the highest attraction and join it. $r_{i,j}$ is originally the absolute power calculated from RSSI between node i and j. Since if the RSSI is higher than -60 dBm, the link is already reliable enough, we fix the RSSI value higher than -60 dBm at -60 dBm to avoid ignoring hop count and path load due to redundant RSSI. M_j is the path load on the node j, which means how many routings are forwarded through the node j. T_j is the number of hops from node j to the gateway. Considering the above three factors at the same time can make the routing algorithm more reasonable and reliable.

4.3 LoRa Ranging and Localization

This section introduces how to use the LoRa Mesh network to locate and track the target. Firstly, we introduce the working flow of location, then process the ranging results of anchors and the target, and finally, give the algorithm of multi-point location.

4.3.1 Localization Workflow

It is important to note that standard LoRa hardware has only one RF channel. To carry out a ranging operation, the packet type parameter of SX1280 should be set from LoRa to LoRa Ranging. This means nodes cannot communicate during ranging. If the workflow of the system is not correct, some nodes in a path that is transmitting data may be in ranging mode, which will cause the data transfer to fail. Therefore, the positioning workflow of the system is very important.

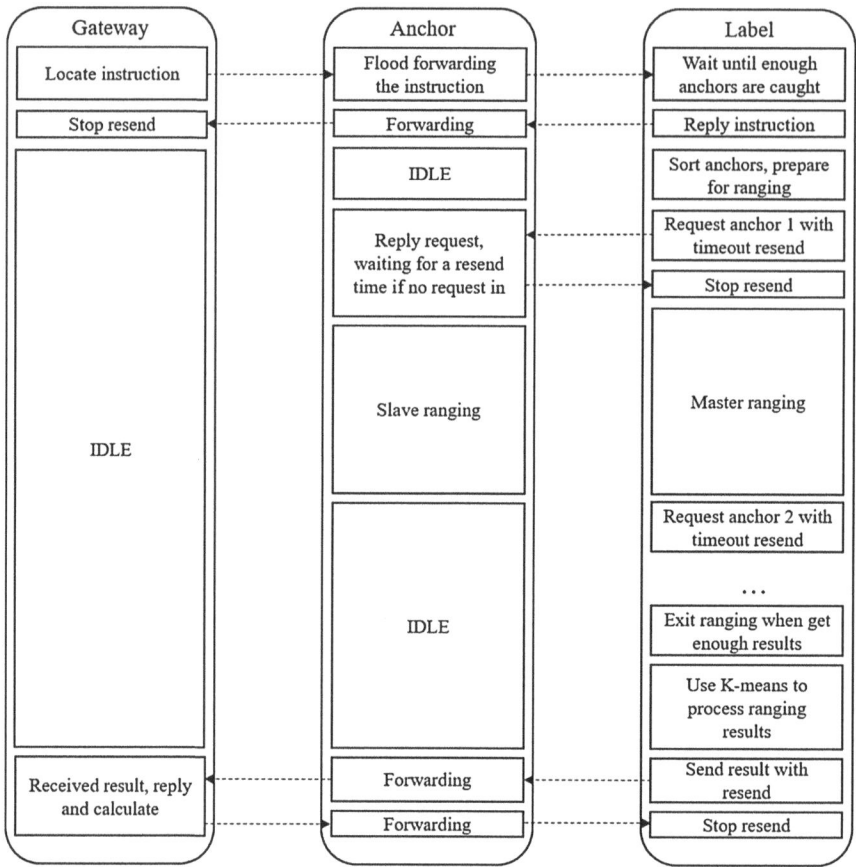

Fig. 4.7 Positioning workflow

We have taken mode switching of LoRa into consideration and designed the workflow of system positioning, as shown in Fig. 4.7. The gateway sends the positioning instruction, and the anchor points in the network broadcast forward the positioning instruction after receiving it. The target is usually in a standby state to save energy consumption. After receiving the positioning instruction, it will record the anchor information of the surrounding forwarding positioning instruction and apply it for joining the network. In the next stage, the target initiates ranging to the surrounding anchor points, until all the surrounding anchor points complete ranging, all the results are sent to the gateway, and the target returns to the standby mode.

Since the target may be moving, they need to find the anchor node currently around before each ranging. Therefore, we broadcast location instructions to the node. When the target receives a broadcast location instruction for the first time, it updates the routing table and collects all broadcast instruction packets from the neighbor node. If the number of nodes around is enough to get locate result, the label responds to the gateway and starts its ranging work.

Figure 4.7 also shows that, during ranging, the target acts as the master role, and the other anchors act as the slave role. This is because only the target knows what the surrounding anchors are. The target records the corresponding RSSI value when receiving broadcast packets from surrounding anchors and sorts the anchors in order of RSSI value. The target will preferentially range anchors with good signal quality (high RSSI value). At the beginning of ranging with one anchor, the target sends a ranging request to the anchor and waits for the reply. If no reply is received within the resend interval time, the target will resend the packet. For anchors, if it receives a ranging request from the target and is ready for the ranging, it replies to the target directly and waits for a resend interval time to listen if a ranging request comes again, which avoids the loss of reply. If the target receives the right reply to the request, it switches to the ranging master mode and will take an extra resend interval time for the anchor to switch to the ranging slave mode. Both the master mode and slave mode have a timeout exit mechanism if no action takes place.

For anchors not near the target, communication or positioning of other targets can be carried out as usual, so that the whole system can carry out both communication and positioning tasks at the same time.

4.3.2 Ranging Algorithm

Because the positioning function of LoRaWAPS is based on the ranging result of anchors and targets. Therefore, it is very important to process the ranging data to make the distance estimation more accurate.

The Semtech SX1280 transceiver has a built-in ranging engine, which performs time-of-flight measurements between a pair of transceiver radios. The distance between the master node and the slave node can be calculated according to the result of the timer:

$$d = N \frac{c}{2^{12} BW} \tag{4.3}$$

In the equation, N is the result of the timer, BW is the transmission bandwidth in hertz, and c is the speed of light.

We first perform ranging operation and ranging correction according to Semtech SX1280 datasheet [1]. To improve distance measurement accuracy, ranging is carried out on 40 different channels successively, and timeout retransmission and exit mechanism are used to cover as many channels as possible. All the results are then processed.

We obtained ToF and RSSI for 40 channels. First, we filter the ToF data to eliminate the extreme cases in the data. Then the RSSI data is filtered and the corresponding data of the channel with large signal attenuation is removed. Secondly, considering the influence of signal refraction and other non-Gaussian noises, we propose to use a K-means clustering algorithm to divide Non Line of Sight (NLOS) and Line of Sight (LOS) results. K-means is a clustering algorithm based on the division of sample sets. We cluster the data of the remaining channels into multiple clusters, in which each cluster corresponds to the data of different channels. The center point of the cluster corresponding to the LOS channel has the minimum TOF value. We then average the results classified as LOS channels to get the output. Finally, because the drift of LoRa hardware clock will lead to the deviation of frequency, some nonlinear results will appear in the close-range measurement, leading to a decrease in the accuracy of distance estimation, which needs to be corrected [2]. The ranging algorithm is described below:

- Step 1: The data of the t-th channel is defined as: $p_t = (d_t, r_t)$, d_t is the TOF ranging data calculated according to Eq. 4.3, and r_t is RSSI value. $\Gamma = \left\{ p_1^{(\Gamma)}, p_2^{(\Gamma)}, \ldots, p_{N(\Gamma)}^{(\Gamma)} \right\}$ represents a set of data and $p_n^{(\Gamma)}$ is an element of the set. Filter H is used for data line filtering, and the remaining data is $\Gamma' \leftarrow H\Gamma$. Filter H can be a median filter, mean filter, or other similar filters.

- Step 2: K-means clustering was used to divide the data in the dataset Γ' into M classes. C_k represents a cluster resulting from k-means clustering. p_c^k represents the center point of the cluster C_k.

$$p_c^k = E(C_k) \qquad (4.4)$$

- Step 3: d_c^k and r_c^k are the TOF ranging data and the RSSI value in the center point p_c^k. Find the cluster which has the minimum ranging distance d_c^k, and the data corresponding to the line-of-sight channel can be extracted.

$$\hat{d} = \min_k d_c^k \qquad (4.5)$$

$\hat{d}$ is the estimated distance. The RSSI value of the center point corresponding to the data of the line-of-sight channel is denoted as $\hat{r}$.

- Step 4: Finally, the estimated distance is modified to get the final result $\tilde{d}$.

$$\tilde{d} = \begin{cases} \hat{d}, & \hat{d} > 100\,\text{m} \\ \gamma\hat{d} + (1-\gamma)10^{\left(\frac{r_0 - \hat{r}}{-10n}\right)}, & \hat{d} \leq 100\,\text{m} \end{cases} \qquad (4.6)$$

Among them, γ is selected by experimental analysis, n is the road loss parameter, and r_0 represents the RSSI value at a distance of 1 m.

4.3.3 Localization Algorithm

With multiple distance information from moving targets to stationary anchors with known coordinates, we are able to solve the coordinates target. The classic three-point positioning method is usually adopted. Assume that the coordinates of the three selected anchor points are (x_a, y_a), (x_b, y_b), (x_c, y_c), the ranging results between the LoRa tag and the three anchors are $(\tilde{d}_a, \tilde{d}_b, \tilde{d}_c)$, and $(\hat{x}, \hat{y})$ represents the estimated LoRa tag position coordinates. The optimal $(\hat{x}, \hat{y})$ is obtained by an iterative method, which minimizes the ranging error $f(x, y) = \sum_{i=a,b,c} \left(\tilde{d}_i - \sqrt{(x_i - \hat{x})^2 + (y_i - \hat{y})^2} \right)^2$. The expression of the $k + 1$ iteration is written as:

$$
\begin{cases}
\hat{x}_{k+1} = \hat{x}_k - f\left(\hat{x}_k, \hat{y}_k\right) \cdot \left(\frac{\partial f(x,y)}{\partial x} \Big|_{x=\hat{x}_k, y=\hat{y}_k} \right)^{-1} \\
\hat{y}_{k+1} = \hat{y}_k - f\left(\hat{x}_k, \hat{y}_k\right) \cdot \left(\frac{\partial f(x,y)}{\partial y} \Big|_{x=\hat{x}_k, y=\hat{y}_k} \right)^{-1}
\end{cases}
\tag{4.7}
$$

Since the real position of the target is unknown, it is impossible to know the positioning error after positioning, so we cannot directly optimize the positioning performance. However, since we know the RSSI measurements of all anchors to the target and the estimated location of the target, we can heuristically estimate the performance of positioning by comparing the difference between measured and calculated RSSI using the estimated location of the target. Therefore, the Posterior RSSI Error (PRE) between the anchor and the target is defined to evaluate the anchor positioning capability.

$$
PRE_i = \sum (r'_i - \hat{r}_i)
\tag{4.8}
$$

Where $r'_i = r_0 - 10\alpha \log_{10} \hat{d}_i$ denotes the wireless signal strength calculated according to the estimated position; r_0 is the radio signal intensity at an interval of 1 m. Based on the above heuristic, the location algorithm is described below:

- Step 1: M different sets are selected from the ranging data of N anchor points to the target obtained, and the location of anchor points $(\hat{x}^M, \hat{y}^M)$ is estimated according to the method in Eq. 4.7.
- Step 2: Calculate the mean PRE of each anchor point corresponding to each $(\hat{x}^m, \hat{y}^m)$ according to Eq. 4.8, $m \in M$.
- Step 3: $(\hat{x}^m, \hat{y}^m)$ corresponding to the set m with the minimum mean PRE is selected as the positioning result.

4.4 Implementation

4.4.1 Control and Visualization Interface

For the control of the system, we designed a control interface, as shown in Fig. 4.8. We can connect the Host computer running our Python script to an arbitrary node in

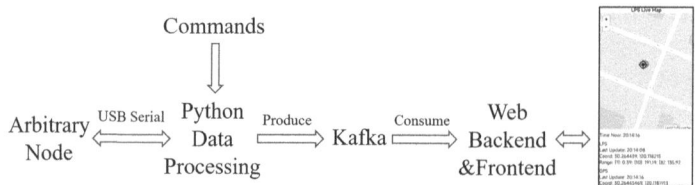

Fig. 4.8 Control and visualization interface

the mesh network through a USB serial. When the Python script receives an input command, it will process it and send further commands to the node. The node will unicast or broadcast the command according to the command type, so the command can be received by one or all nodes in the mesh network. Replies received by the node will be sent back to a host computer. In particular, localization results are produced to a Kafka topic data stream. A web page will consume that data and visualize them on a map. GPS data for experiment use can also be collected from an Android device with a custom app and uploaded to be shown on the map.

4.4.2 Anchors Deployment

We plan to deploy the system in Yuquan Campus of Zhejiang University. In the actual deployment of the system, we should consider the selection of the most appropriate anchor location to better cover the positioning blind area. With the help of electromagnetic simulation software, the signal intensity of anchors at different positions can be generated to help evaluate the coverage ability of anchors.

The actual deployment requires the coverage area to receive stable signals from at least three anchor points ($RSSI \geq -85$ dBm). Through the heuristic algorithm, we hope to use as few anchor points as possible to provide positioning ability covering Yuquan Campus. The algorithm used is as follows:

- Step 1: As shown in Fig. 4.9a, we assume several locations l_n where anchors can be placed at intervals of 5 m. Then we converted the spatial grid into grids spaced 1 m apart and calculated the signal intensity of the anchor placed at position L_n on all grid points, denoted as S^n. The member s_j^n of the set S^n denotes the signal strength of the anchor placed at the n-th point at the j-th lattice point. Its value is obtained by electromagnetic simulation software. At this point, each position l_n in the set L can place anchors. c_j represents the signal coverage quality at the j-th lattice point, which is 0 data when there is no signal.
- Step 2: Suppose that after placing an anchor point at l_n,

$$c_j^{(n)} = \begin{cases} c_j + 1, s_j^n > -85 \text{ dBm} \quad \& \quad c_j < 3 \\ c_j, else \end{cases} \tag{4.9}$$

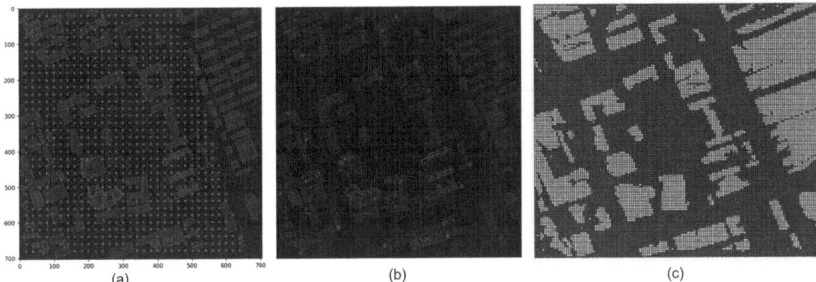

Fig. 4.9 Anchor deployment location selection and area coverage capability. (**a**) Alternative deployment locations. (**b**) Optimal placement of anchors. (**c**) Location capability coverage

- Step 3: Find n_0 for which $C^{(n)}$ is the largest,

$$C^{(n)} = \sum_j (c_j^{(n)} - c_j).\qquad(4.10)$$

Move n_0 out of set L and add it to set $L\prime$, whose elements indicate where to place the anchor.

$$c_j = c_j^{(n_0)}.\qquad(4.11)$$

- Step 4: Repeat steps 2–3 until you find a set $L\prime$ of k elements, set $L\prime$ represents the lattice point where the optimal k anchor points are placed.

Based on the above algorithm, we carried out numerical experiments in Yuquan Campus of Zhejiang University. Firstly, a number of green grid points 5 m apart are marked on the map to indicate the alternative anchor position, as shown in Fig. 4.9a. We found that only 18 anchor points could be deployed to provide a stable location service capability for the entire outdoor area of the campus. The green dots in Fig. 4.9b indicate where the anchor points should be placed. Figure 4.9c shows the signal coverage capability when 18 anchor points are placed. The blue area in the figure indicates that positioning signals from more than three anchor points can be received. The white area that cannot provide positioning services basically corresponds to the building area.

4.5 Evaluation

This section reports on how our system performs in a real-world environment. Section 4.5.1 demonstrates the ranging performance of LoRaWAPS. Section 4.5.2 demonstrates the positioning performance of LoRaWAPS.

4.5.1 Ranging Experiment

We first evaluate the transmission capability of the LoRa signal. We conducted outdoor ranging experiments using standard LoRa hardware, the specifications of which are described in Sect. 4.1. The LoRa hardware we used tried 40 channels per ranging, and not every channel was successful. Recording the number of successful ranging channels at different distances can evaluate the quality of the signal and the effect of ranging. We also recorded the RSSI of each experiment and calculated the ranging error from the ranging result and the GPS result.

In the experiment, the working mode of master-slave ToF ranging was adopted, and the spread factor SF = 6. The transmission power is 20 mW. This experiment was carried out between two nodes on a straight and relatively clear road, ranging was performed approximately every 25 m, the result is then filtered and evaluated, shown in Fig. 4.10.

In the figure, the blue lines represent the RSSI of the received signal, decreasing gradually as the distance increases. The green bar chart shows the number of successful ranging channels, which also decreases gradually as the distance increases. The yellow line indicates the error of ranging. As the distance increases and the signal quality decreases, the ranging error also increases gradually. RSSI and ranging errors are not only affected by distance, but also by non-Gaussian noise such as multipath, so the curve in the figure is not monotonically changing. All in all, LoRa's ranging ability is strong, with the longest ranging distance reaching over 1000 m. The ranging error is below 15 m. In particular, LoRa's ranging capability is excellent up to 500 m without shading, which provides a good basis for location services.

Next, we evaluate the proposed ranging algorithm. Other mainstream LoRa systems were compared in the experiment [2, 3]. In the experiment, the LoRa anchor was placed high in the center of a square area with a side length of 200 m. The target

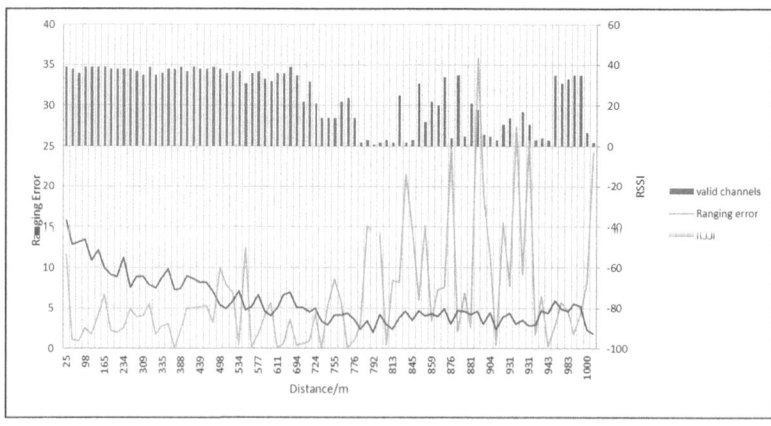

Fig. 4.10 ToF performance test

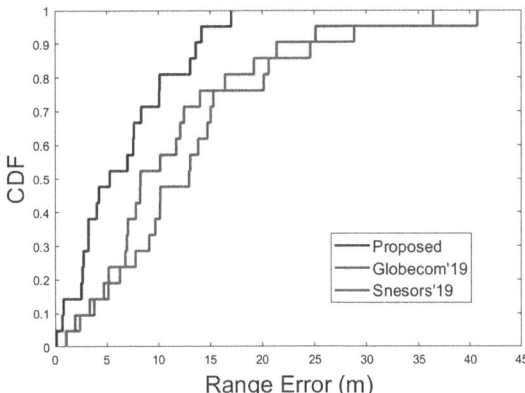

Fig. 4.11 CDF of range algorithm

was measured at 30 different locations in the area. And record the error between the ranging results and the actual distance in each experiment. The cumulative distribution function of ranging errors is shown in Fig. 4.11.

In the figure, the blue lines represent our proposed ranging method with an average error of 6.46 m, 75% of which is less than 10 m. The other two mainstream LoRa systems have ranging errors of 11.48 and 13.58 m, respectively. They use a mean filter or Kalman filter to reduce the influence of noise on the ranging results, but it can not overcome the interference of non-Gaussian noise caused by the multipath effect. The proposed method overcomes the influence of the multipath effect by clustering on the basis of filtering and is proved to be effective by experiments.

4.5.2   Positioning Experiment

Based on the previous steps, this section evaluates the positioning performance of the system.

In the experiment, a total of 9 LoRa anchor points were placed in a square area with a side length of 500 m. The location of the anchor points was determined according to the anchor point deployment algorithm described above. 9 anchor points were networked by wireless mesh, 30 different positions of the target were located in the region and the positioning errors were recorded. The bandwidth, spread factor, and other parameters of the system are the same as those of the ranging experiment. We compare the mainstream LoRa positioning system, which uses all anchor points for position calculation [4]. The cumulative distribution function of positioning errors is shown in Fig. 4.12.

In the figure, the red line represents the positioning error of LoRaWAPS, with an average error of 14.49 m and 70% of the errors being less than 20 m. The blue line corresponds to the positioning error of [4], with an average positioning error

Fig. 4.12 CDF of
localization algorithm

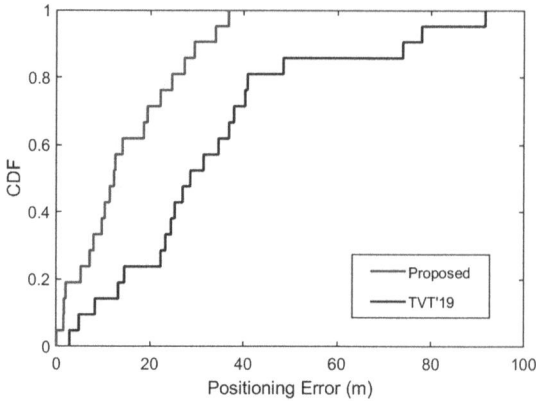

Fig. 4.13 Influence of
anchor density on positioning
ability

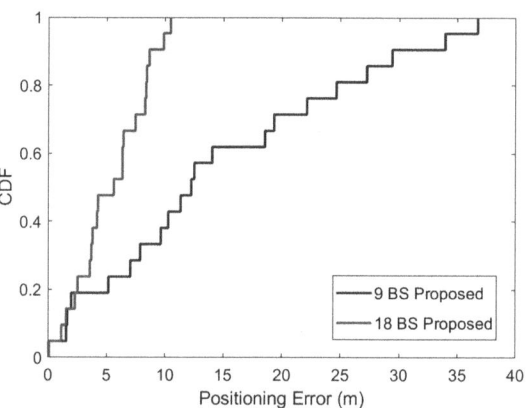

of 33.93 m. Thanks to the improvement of the ranging algorithm and positioning algorithm, LoRaWAPS shows better positioning performance. However, we found that due to the occlusion of buildings, trees, and vehicles, there is still quite a large area of poor signal quality, and the ranging and positioning accuracy in these areas are poor, which leads to the average positioning error of the system is greater than 10 m.

To investigate the effect of the density of LoRa anchor points on positioning performance in the region, we deployed 18 anchor points in the same square area with a side length of 500 m to achieve better signal coverage. All other Settings are the same as in the previous experiment. The cumulative distribution function of positioning error for different anchor point densities is shown in Fig. 4.13.

In the figure, the red lines represent the positioning errors of the system when 18 anchor points are deployed. The average positioning error is less than 5 m, and the maximum positioning error is less than 10 m. Considering that the positioning error of civil GPS as a reference is about 5 m, it can be considered that the system has positioning capability close to GPS when the deployment density of anchor points is high.

Finally, we conducted system-level experiments to prove that all parts of the system can run smoothly. We have deployed LoRaWAPS in Yuquan Campus of Zhejiang University, with a total of 21 anchor points, which can cover most areas within the campus. In order to evaluate the real-time positioning accuracy of the LoRaWAPS, the presenter moved freely within the campus with a target node and a mobile phone with satellite positioning capability. The mobile phone records the GPS coordinates in real time and uploads them to the server. LoRaWAPS locates the target node through anchor points and uploads the location results to the server as well. The presenter can access the server via a web page to compare the GPS positioning results with LoRaWAPS positioning results in real time. The web page is shown in Fig. 4.14.

In the figure, blue circles represent LoRaWAPS location results and black circles represent GPS location results. At the bottom of the map are displayed successively in real time: time, position coordinates calculated by LoRaWAPS, anchor numbers involved in positioning in LoRaWAPS, ranging results of each anchor, and position coordinates provided by GPS. It can be seen that in most cases, the positioning error of LoRaWAPS is within 15 m. This accuracy is acceptable for most IoT location scenarios. If better positioning accuracy is needed, simply increase the density of anchor deployment in exchange for better signal coverage.

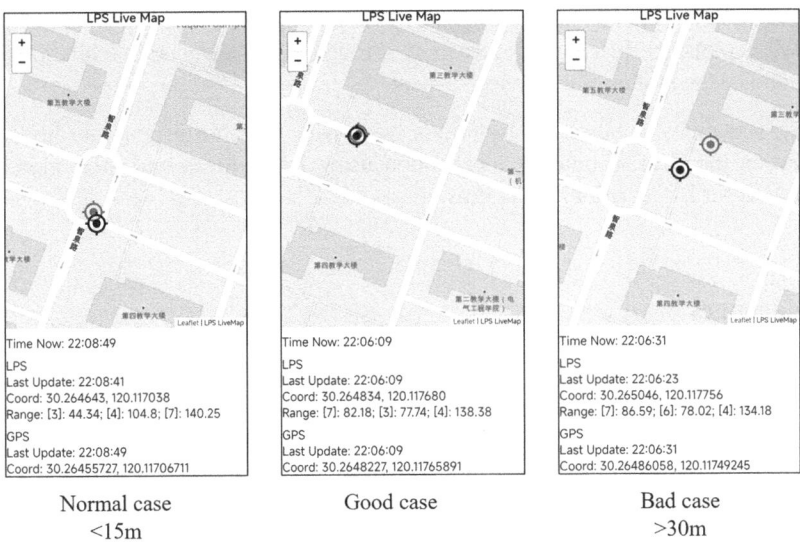

Normal case   Good case   Bad case
<15m   >30m

Fig. 4.14 Localization system test

4.6 Summary

In this chapter, we explore outdoor localization using multiple interconnected base stations through LoRa Mesh. We present a cost-effective, low-power, and easily deployable IoT positioning system. The key contributions of this chapter can be summarized as follows:

- Designing a LoRa Mesh networking protocol: We introduce a protocol that incorporates channel activity detection and packet CRC check mechanisms. These optimizations aim to reduce packet congestion and enhance communication reliability.
- Proposing a routing algorithm: Our algorithm considers factors such as communication delay, communication reliability, and node load. It has been proven to be more effective than existing routing algorithms in terms of both delay and communication quality.
- Designing a positioning method based on LoRa Mesh: We present the workflow of LoRa communication and the positioning system for the first time. Additionally, we build a LoRa communication-positioning perception system.
- Introducing a clustering-based ranging algorithm and anchor point selection-position solution algorithm: To address the challenges posed by NLoS diameter and noise on LoRa ranging accuracy, we propose these algorithms. Experimental results demonstrate their superior accuracy compared to existing three-point positioning algorithms.
- Discussing optimal anchor placement based on electromagnetic simulation: We explore the placement of anchors through electromagnetic simulation and validate the system's actual deployment.

By accomplishing these contributions, we provide valuable insights and advancements in the field of outdoor localization using LoRa Mesh and offer practical solutions for IoT positioning systems.

References

1. Semtech. Sx1280, Accessed: 2021 [Online]. Available: https://www.semtech.com/products/wireless-rf/24-ghz-transceivers/sx1280
2. K. Hu, Y. Chen, S. He, Z. Shi, J. Chen, Z. Tao, iLoc: a low-cost low-power outdoor localization system for Internet of Things, in *2019 IEEE Global Communications Conference (GLOBECOM)*, Hawaii (2019), pp. 1–6
3. P. Savazzi, E. Goldoni, A. Vizziello, L. Favalli, P. Gamba, A Wiener-based RSSI localization algorithm exploiting modulation diversity in LoRa networks. IEEE Sens. J. **19**(24), 12381–12388 (2019)
4. K.-H. Lam, C.-C. Cheung, W.-C. Lee, RSSI-based LoRa localization systems for large-scale indoor and outdoor environments. IEEE Trans. Veh. Technol. **68**(12), 11778–11791 (2019)

Chapter 5
Enable Angle of Arrival in LoRa for Efficient Indoor Localization

Abstract Though angle of arrival (AoA) has been recently adopted in many localization systems (e.g., WiFi, Bluetooth), it is still an open issue whether it can be enabled in LoRa for long-range localization. There are mainly two challenges: (1) current LoRa platform is not equipped with multiple RF channels for the array, and (2) the phase information is unavailable from the PHY layer. In this chapter, we propose a new method utilizing the ranging information for efficient AoA estimation, showing that multi-channel capability and phase information are not necessary. Specifically, we design a mobile gateway with a virtual array extended by an RF switch on a single channel. We further modify the ranging procedure by optimizing the hopping mechanism and packet format. To improve the AoA performance, we design a binary classification framework that achieves a theoretical error of 180/n degrees, where n is the number of antennas. Further, we propose a "rotate and follow" strategy to create more "virtual antennas" for the AoA estimation. The prototype localization system, RLoc, is built for both LoS and NLoS scenarios. Experiments show that RLoc achieves an average AoA estimation error of 2.4 degrees and is efficient in kilometer-level area, which outperforms to the existing localization systems.

Keywords LoRa · Internet of Things · Angle of arrival · Ranging · Localization

5.1 Problem Formulation and Challenges

LoRa implements the ranging by measuring ToF between the master and the target. The measurements easily deviate from the actual distance due to hardware uncertainty and environmental interference. Many previous studies have investigated the long-range localization by its ToF and there is a consensus that localization only based on ToF has a poor performance with a hundred-meter level accuracy in wide-area outdoor environments [1]. It performs even worse in indoor environments where signal attenuation and multiple path effects are severer.

In this paper, we take the first attempt to explore the feasibility of enabling AoA estimation in LoRa. We are given a commercial LoRa platform that has been

© The Author(s), under exclusive license to Springer Nature Switzerland AG 2024
Z. Shi et al., *LoRa Localization*, SpringerBriefs in Computer Science,
https://doi.org/10.1007/978-3-031-48008-9_5

deployed in numerous applications and needs to build a gateway as the master node for the localization of the commercial LoRa systems. The gateway initiates the ranging procedure by sending a ranging packet. Then, the target node receives the packet and replies with a response ranging packet. During repeated ranging procedures, the gateway estimates the AoA of the target.

We note that it is challenging to enable AoA estimation in commercial LoRa platforms due to the following reasons:

- It is infeasible to extract the CSI information and the phase information from current LoRa platforms, and thus phase difference is not available. Are there any other measurements that can be taken by two different antennas and are used for efficient AoA estimation?
- Due to the low-cost and low-power design, the LoRa chip only has one RF channel. Even though we have an antenna array connected to the RF channel, only one antenna can receive and process the received signals at the same time. How to reduce the effect of the absence of multiple channel chains and improve ranging difference estimation using multiple antennas?
- As shown in Sect. 5.3, the classic approach to AoA estimation results in poor performance in RLoc when phase information is not available. Is there any new approach that has low computational complexity and high accuracy?

5.2 Redesign of LoRa Gateway and Ranging Procedure

In this section, we propose our solution to AoA estimation in LoRa. We will first redesign the gateway and then the ranging procedure for AoA estimation.

5.2.1 Gateway Redesign

The LoRa gateway is desired to be light and small such that it can be deployed for most mobile applications. We simplify the structure of a commercial gateway and only keep basic components for ranging, controlling, and communication. The designed gateway is comprised of a LoRa chip (SX1280), an MCU (STM32L476), an RF switch (HMC241), a BLE module (CC2640), and a lithium battery (600 mAh). Since there is only one RF channel integrated into the LoRa chip, we redesign the gateway with an RF switch, which is in charge of antenna switching between the only RF channel and multiple antennas. To reduce the antenna switching time, we select a single pole 4 throws (SP4T) chip HMC241, which is able to switch between two antennas within 150 ns [2]. Further, the gateway integrates a low-power BLE module to connect to a smartphone that is used for obtaining gateway orientation and AoA computing. Note that the inclusion of a smartphone is not restrictive and can be replaced in practical applications. The simplified structure significantly reduces the weight of the gateway to 112 g and the size to 13 × 13 cm.

We design a circular antenna array connected to the RF switch that can reduce the size of the gateway. Antennas uniformly distributed on the circular array avoid the blind area and increase system coverage. Further, the circular shape of the antenna array can be exploited to enhance the AoA estimation accuracy, that is, we can rotate the circular array to increase the number of virtual antennas which can reduce the estimation error of the AoA (see Sect. 5.3).

5.2.2 Leveraging Ranging Difference

Compared with MIMO systems (NI USRP RIO, Intel 5300 NIC), it is unavailable to utilize the phase difference of the same packet that is transmitted by the target and received at different antennas. We propose to exploit the ranging difference to replace the phase difference in LoRa. Here, it means the difference in the time of flight of the signals received at two antennas. Clearly, the measurement can be used to calculate the distance difference from the target node to these two antennas. Denote the number of antennas by n, the angle between antenna i and antenna 1 by α_i, and the radius of the circular antenna array by r (see Fig. 5.1). For the antenna i and antenna 1, the distance difference, calculated by Δd_{i1}, is

$$\Delta d_{i1} = c \times (ToF_i - ToF_1), i = 2, 3, \cdots, n. \tag{5.1}$$

Δd is correlated with AoA θ as follows:

$$\Delta d = \begin{bmatrix} 2r\sin(\frac{\alpha_1}{2})\sin(\theta - \frac{\alpha_1}{2}) \\ 2r\sin(\frac{\alpha_2}{2})\sin(\theta - \frac{\alpha_2}{2}) \\ \cdots \\ 2r\sin(\frac{\alpha_n}{2})\sin(\theta - \frac{\alpha_n}{2}) \end{bmatrix}. \tag{5.2}$$

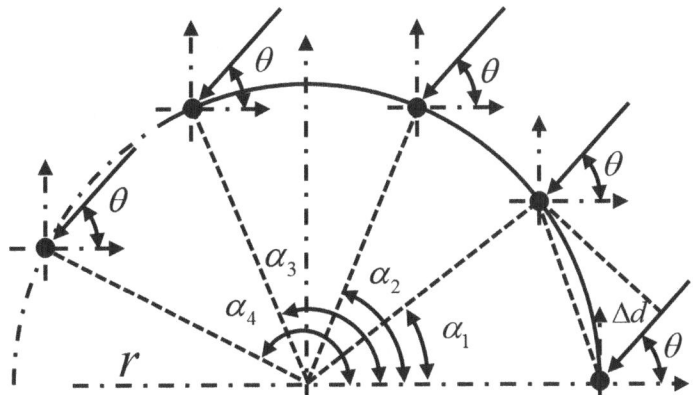

Fig. 5.1 Antenna array model

Therefore, we will be able to obtain the estimation of AoA when Δd measurement is accurate. However, it is challenging to obtain an accurate estimation of Δd due to the following reason. (1) The best ranging resolution is more than 1/3 of the antenna space so that the Δd accuracy is limited. (2) Though we install an antenna array on the gateway, two antennas can not receive and process the same response packet from the target node. This is because commercial LoRa chip only has one RF channel, which can be connected to only one antenna at a time. As a result, each antenna needs to sequentially measure the ToF, which introduces the noises to the difference estimation (see Fig. 5.5 in Sect. 5.5.1).

To address those issues, we first need to reduce the time interval between which two ranging packets are transmitted using the same channel. At the best case, if the time interval approaches 0, we can take it as that two antennas receive the same ranging packet. Hence, the uncertainty and interference can be removed from the difference. Unfortunately, the ranging procedure in LoRa is not designed for such a purpose and has a poor performance for the difference estimation.

5.2.3 Ranging Procedure Redesign

In this subsection, we redesign the ranging procedure that contains the ranging packet format, hopping mechanism, and ranging preprocessing. The time for transmitting and receiving one ranging packet at each subchannel is mainly comprised of (1) transmission time, (2) propagation time, and (3) processing time. The propagation time is determined by the distance and the signal propagation speed (i.e., the speed of light) which are fixed. We focus on the optimization of transmission time and processing time. A ranging frame contains the preamble, headers, and ranging symbols. Clearly, a short frame helps save transmission time. Thus, we configure the minimal ranging symbols and preamble to decrease the length of the ranging packet in the physical layer. Specifically, we modify the parameters of the ranging engine in the firmware to form a new frame for difference estimation. Resultantly, the ranging frame has now less than 61 symbols as shown in Fig. 5.2. The transmission time is reduced to 1.19 ms which may take more than 300 ms for each packet in the standard raging [3].

Further, in the application layer, to construct the complete packet, the controller needs to encode the frame parameters such as packet type, coding rate, and

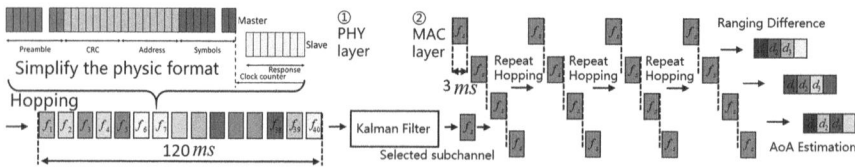

Fig. 5.2 Ranging procedure redesign

bandwidth before the transmission, which increases the processing time. Notice that we focus on the localization applications and these parameters are constant. We can compute all the ranging packets in advance during the idle status so that they can be sent to the modem directly. To this end, we move the communication packet calculation from the hopping to the system initialization, which saves 13 commands in the ranging and 17 commands in the hopping process. Such a design can reduce 30 commands and save 0.02 ms in each hopping channel. In the whole hopping procedure, it helps reduce 3.2 ms of the processing time.

In the ranging procedure, LoRa hops to 40 subchannels from 2.40 to 2.48 GHz to obtain a stable estimation of distance, i.e., the distance is estimated as a median of measurements of 40 subchannels. For AoA estimation, if we execute the ranging procedure sequentially and traverse all the 40 subchannels, it will take 120 ms for one round of frequency hopping and thus the time interval between two ToF measurements can be as long as 240 ms. Hence, we optimize the hopping process by introducing an antenna switching mechanism in the ranging procedure. This is achieved by inserting an antenna switching in the selected subchannel ranging. Specifically, RLoc first completes a full hopping for all subchannels to measure the distance from the target. The measurements are calculated via a Kalman filter as the result of ranging. When it starts for AoA estimation, the subchannel is fixed at the selected one that is the closest to the ranging result. Then, RLoc switches the connected antenna between the ranging. Experimentally, the RF switch takes less than 1 μs which is negligible. When the best ranging subchannel changes, it is forwarded to the AoA process periodically. As illustrated in Fig. 5.2, we add a switching task in the hopping process periodically and this greatly reduces the time interval from 120 to 3 ms for two antennas to receive the ranging packet on the same subchannel.

To mitigate the interference and uncertainty in the ranging, the LoRa engine calibrates the fixed and design-specific delays in the hardware, which contains 35 commands. Notice that we concentrate on calculating the difference in a short time for AoA estimation. The fixed and design-specific delay is constant in two adjacent ranging procedures and can be ignored in the gateway. Thus, we only process the calibration at the first of AoA estimation and remove those commands during the ranging procedure to reduce the time interval.

5.3 Improving AoA Estimation via Rotation

When the difference is measured, the AoA, denoted by θ, is estimated via (5.2). We conduct experiments to evaluate the AoA estimation performance based on the difference directly.

The results are shown in Table 5.1.

The error is defined as the absolute bias between the estimation and ground truth in Table 5.1. Clearly, the estimation errors in most sectors are too prominent to be ignored. The experimental results demonstrate that the traditional scheme has

Table 5.1 Experimental results of AoA estimation (unit: degree)

Angle	15	30	45	60	75	90	105	120
Error	48	27	22	17	11	4	14	9
Angle	135	150	165	180	195	210	225	240
Error	22	37	63	98	53	18	21	24
Angle	255	270	285	300	315	330	345	360
Error	21	7	14	18	16	33	67	55

a poor performance in LoRa AoA estimation since it needs accurate information of ranging differences while due to hardware constraint, we only obtain the noisy measurements.

5.3.1   Binary AoA Classification

In this subsection, we present an efficient scheme for AoA estimation.

We take an in-depth investigation of the experimental results of the difference measurements. For two antennas, the line is called the segmentation line of AoA plane that is perpendicular to the line segment connecting these two antennas. It is clear that such a segmentation line of AoA plane divides the possible AoA space into two half planes. We obtain a stable interesting finding from the experimental results: most of the time, the values of the difference are greater than zero when the actual AoA is in the right half plane and it is negative in the left half plane. With such a finding, it is possible to make a stable decision about which half plane the AoA is in. We model the AoA estimation as a binary classification problem in the following to improve the stability.

Now we focus on the classification accuracy bound with n antennas. Connecting two adjacent antennas in the circular array constructs an equilateral polygon with n vertices. Since one antenna can pair with any other antenna in the array, the number k of antennas (i.e., vertices) that lie in the inferior arc between the two antennas (i, n) on the circle varies from 0 to $n - 2$. When k is an even, the side $l_{i,n}$ connecting the pair of these two antennas is parallel to the side of two middle vertices. If k is odd, the side $l_{i,n}$ is parallel to one side of the equilateral polygon. Therefore, we only need to consider pairs of two antennas with three antennas in-between them as illustrated in Fig. 5.3a, which simplifies the analysis. The angle β between those two perpendicular bisectors which are the boundary of the classification area demonstrates the estimation error of the classification scheme, that is

$$D_{accuracy} = \beta = \frac{180°}{n}, \quad n \geq 3. \tag{5.3}$$

Obviously, the estimation error bound of the binary classification is low when a small number of antennas are used. For example, when the gateway is equipped

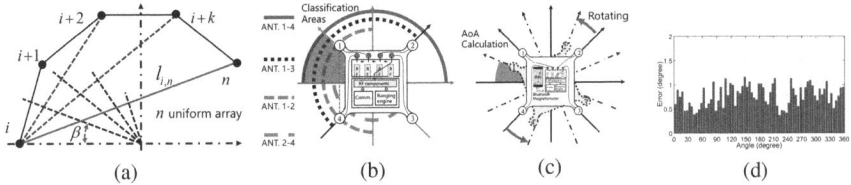

Fig. 5.3 Adding "virtual antenna arrays" for AoA estimation. (**a**) Resolution of n uniform antenna. (**b**) AoA classification. (**c**) Rotate to localize. (**d**) IMU orientation estimation

with 4 uniform antennas, the AoA estimation error is 45°. An illustration is shown in Fig. 5.3b. To improve the AoA accuracy bound $D_{accuracy}$, we need to put more antennas on the circular array. However, the hardware constraint (e.g., antenna size) and manufacturing cost make it impossible to deploy too many antennas. In this paper, we propose a novel way of adding "virtual array" by rotating the circular array during estimation to improve the accuracy, which will be discussed in the next subsection.

5.3.2 Virtual Array

The key idea of adding "virtual arrays" to achieve better performance is to change the orientation of the circular array continuously to establish a collection of virtual circular antenna arrays. During the rotation, the gateway keeps sending and receiving the ranging packets, and the orientation of the antenna array is also recorded. According to Eq. 5.3, the classification accuracy increases as we keep rotating the circular antenna array to simulate more antennas.

We proceed to elaborate on the implementation of the virtual array using the gateway with a circular antenna array. Since the gateway is co-located with a smartphone, we exploit sensors embedded in the smartphone to record the orientation of the gateway during rotation. Without the integral error, the magnetic field is stable and can be estimated from the magnetometer. However, the attitude of the smartphone is uncertain during rotation, leading to an estimation bias of the orientation. To calibrate the magnetic field measurements, we utilize the accelerometer and gyroscopes to estimate the attitude of the gateway and then fuse them with the magnetometer to correct the array orientation via a particle filter. As a result, the orientation can be calculated by

$$\alpha(t) = hN(t)M(t), \tag{5.4}$$

where h denotes the projection matrix which transforms the orientation vector to the horizontal plane, N indicates the IMU measurements, and M is the magnetic field vector. We conduct experiments on a commercial smartphone to analyze the orientation estimation and the results are shown in Fig. 5.3d. We can see that the orientation estimation error is less than $1.5°$ and the average is $0.77°$. This shows the feasibility of rotating the circular array in practice to increase more virtual arrays.

5.3.3 AoA Calculation

Although numerous virtual arrays are deployed in the previous subsection, they are connected during the random rotating when the virtual arrays may be not constructed distributedly. We now proceed to calculate a specific AoA utilizing all the ranging difference results. We denote the probability of the AoA by $p(\alpha, t)$ at time t where α is the orientation of the gateway. The starting boundary of the classification area is denoted by β. To alleviate the multi-path reflection, we weigh the difference based on the corresponding RSSI due to the signal-strength loss in the reflection. Hence, the LoS signals from the target are enhanced in the AoA optimization which improves the AoA estimation performance. Specifically, the RSSI weights the probability via $W(\alpha, t)$. Thus, the AoA probability model can be expressed as

$$P_{AoA} = \sum_{i}^{n} \int_{\beta_i}^{\beta_i + \frac{\pi}{4}} p(\alpha_i, t) W(\alpha_i, t) d\beta. \tag{5.5}$$

Without loss of generality, we assume that the probability of the classification area is uniform which can be $p(\alpha, t) = p_0$. Then, the AoA θ can be calculated via solving the optimization problem:

$$\max \ P_{AoA}(\alpha, \beta, t)$$
$$s.t. \quad \begin{cases} \alpha \in \alpha_i & i = 1, 2, 3, \cdots, n. \\ t_0 \leq t \leq t_f \end{cases} \tag{5.6}$$

We can find the optimal β representing the θ when the P_{AoA} is maximized.

5.4   Implementation and Applications

5.4.1 Implementation

We implement the AoA estimation and prototype a system named RLoc, which can perform tracking as well as localization for the target. The gateway is connected through a BLE module to a smartphone that takes measurements of gateway orientation and executes the AoA and localization algorithms as illustrated in Fig. 5.4. Specifically, we develop an application on the Android platform. The IMU

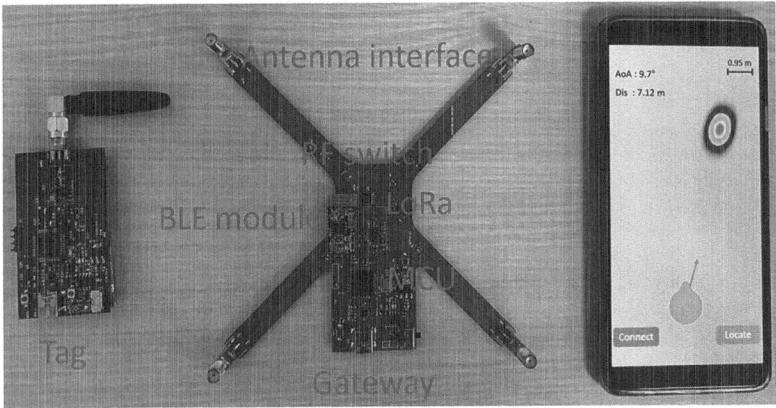

Fig. 5.4 RLoc implementation that contains a tag and a redesigned gateway

and magnetometer data are obtained from the integrated sensors. When the attitude data is updated, we push them into the recording module during the rotation. Then, they are fused with the ranging results obtained from the gateway. A user interface is developed for instructions and visualization in the localization.

5.4.2   Applications

With the capability of AoA estimation, RLoc can support many outdoor and indoor localization applications. We here provide two basic application scenarios: LoS localization and NLoS localization.

LoS Localization In such a scenario, the target node is in direct sight of the gateway. When the gateway is fixed, we can measure AoA as well as the distance associated with the target node. Then we can calculate the relative position of the target node using only one gateway. As aforementioned, the estimation of distance usually contains large noise, leading to a coarse location estimation. When we can move the gateway around, e.g., the gateway is piggybacked on a robot with a rotation controller, we can move the gateway to more than two different locations, and perform localization jointly.

NLoS Localization In this scenario, the target node can not be directly localized. We assume the gateway is mobile (e.g., the gateway is piggybacked on a robot) so that we can iteratively localize the target node by moving the gateway around, following the AoA estimated at the current location. At the initial location, the gateway estimates the AoA by rotating the circular array. Since AoA demonstrates the orientation of the target, the gateway can follow the direction to approach the target node. At each location, the gateway can perform AoA estimation again. The

---

Algorithm 2: Iterative localization algorithm

Input: Target ID
Output: position of the target node S
Wake up the target from sleeping mode;
while *not arrive* **do**
    Rotate RLoc and start ranging;
    Update IMU N and magnetic M estimation to calculate orientation α via (5.4);
    Encode ranging difference measurements with α;
    Calculate the probability p weighted by $W(\alpha,t)$;
    Accumulate the P_{AoA} via (5.5) and estimate the angle θ via (5.6);
    Follow the AoA estimation;
    if *Find the target* **then**
        return the position S;
    end
end

---

process continues iteratively as illustrated in Algorithm 2. The iterative process terminates when the gateway arrives at the target location.

NLoS localization has many applications. For example, in a large factory, there are many valuable assets, which could be placed anywhere and no one knows their actual positions. With LoRa tags attached to these assets, they could be easily tracked. Another interesting application is contactless file delivery. Currently, COVID-19 has spread to most of the countries in the world. To avoid close contact, we can use robots equipped with a gateway to deliver files and meals to the target persons with LoRa tags.

5.5   Performance Evaluation

In this section, we evaluate the accuracy of ranging difference estimation, AoA estimation, and localization performance. The experiments are conducted both indoors and outdoors. In the experiments, the ground truth of the target angle and position are measured via a protractor and a laser rangefinder, respectively. We select the Samsung Galaxy A9 Star as the co-located smartphone that is equipped with an accelerometer, a gyroscope, and a magnetometer.

5.5.1   Ranging Difference Estimation

We first analyze the measurements of ranging differences to evaluate the performance of the binary classification framework. In the experiments, the gateway is fixed and the target tag is placed in front of the two selected antennas at a distance of 10 m. The gateway switches the two antennas to measure the distance

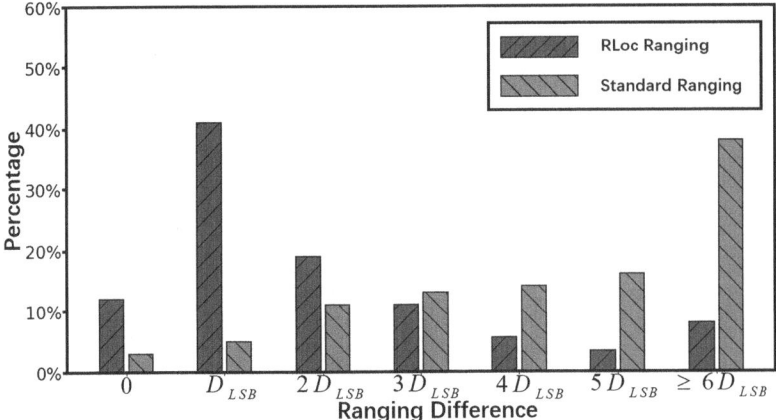

Fig. 5.5 Experimental results of ranging difference

difference 2000 times. For comparison, we also conduct experiments using the standard ranging procedure. The results are depicted in Fig. 5.5.

The ranging difference is the result of round trip distance. Hence, the resolution D_{LSB} in 1.625 MHz is 2.25 cm. We see that by using the standard ranging procedure of LoRa, 79.4% of measurements are greater than 3 times the best resolution. In RLoc, 83.5% results fall in the range of 3 times the best resolution that is close to 1/2 of the wavelength λ. The improvement of ranging difference accuracy increases the stability of the binary classification framework in RLoc.

5.5.2 AoA Estimation

5.5.2.1 AoA Estimation in Different Spreading Factors

The fast antenna switching and accurate distance ranging are two important factors in AoA estimation. In order to obtain a more accurate ranging performance, we need to increase the spreading factor (SF) [4]. However, the big SF takes more hopping time during the switching. We proceed to conduct experiments to evaluate the performance of AoA estimation in different SFs. The target node is deployed on a set of positions at a circle and the gateway is fixed at the center of the circle. The experiments are conducted at a meeting room (7 m × 10 m).

The average AoA estimation error and average distance estimation error are illustrated in Fig. 5.6. The distance error is small when the SF becomes higher. Meanwhile, it doubles the transmission time. Hence, the AoA error increases in high SFs. Determined by the distance and ranging time, RLoc achieves the best performance in SF5 the average error is 2.4 degrees. This is because the improvement in the distance is canceled by the time cost. To our best knowledge, it outperforms the CSI-based system (WiFi) whose accuracy varies from 6° to 15°

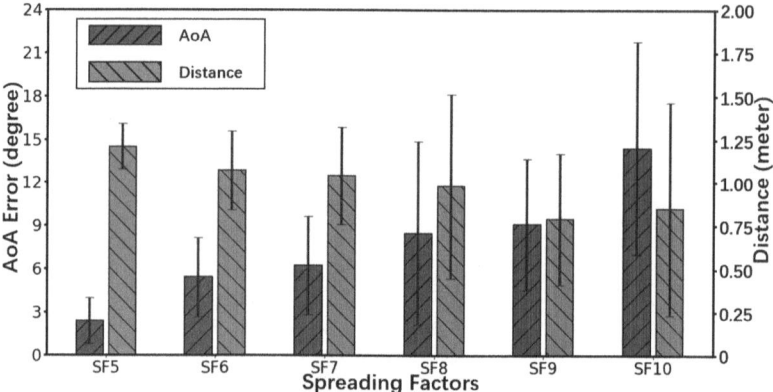

Fig. 5.6 Experimental results on AoA estimation

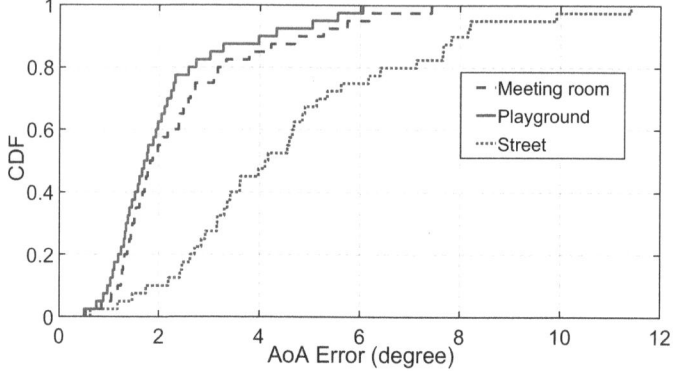

Fig. 5.7 Experimental results on AoA estimation

(e.g., 14° in D-MUSIC [5], 9.1° in Array-Track [6], 6.7° in RoArray [7] and 6.62° in SpotFi [8]).

5.5.2.2 AoA Estimation in Different Environments

To evaluate RLoc robustness in different environments, we conduct experiments in a meeting room, an outdoor playground, and a street, respectively. The SF is configured as SF5. In the street, the target and the gateway are deployed along one side. We start AoA estimation from different orientations for 40 times experiments. The ground truth is recorded via a photoelectric encoder and a laser rangefinder.

As illustrated in Fig. 5.7, the average error in the playground is 2.2 degrees. In an open environment, since there is little interference, the 80% results are smaller than 4 degrees. The average error in the meeting room is 2.4 degrees which is close to

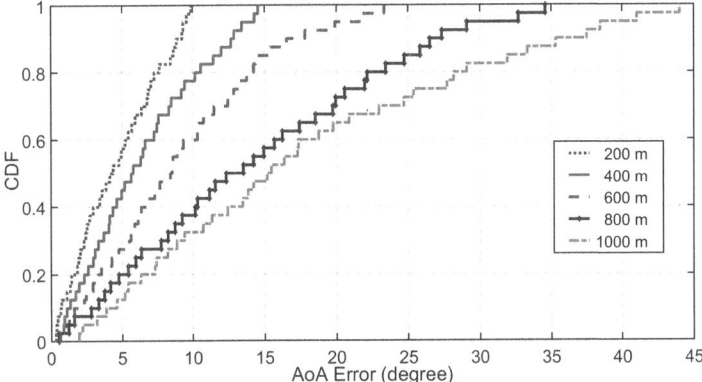

Fig. 5.8 Experimental results on AoA estimation

the result in the playground. When RLoc is deployed in the street, the performance is degraded by an average of 4.6 degrees. Because the trees and cars along the way increase the reflection and the interference. However, the 80% results are less than 8 degrees. Those results demonstrate the stability of RLoc for indoor and outdoor applications.

5.5.2.3 AoA Estimation for Long-Rang Scenario

As long range may lead to more uncertainty, we conduct experiments from 200 to 1000 m to evaluate the accuracy in large scenarios. The target is deployed on the highway at every 200 m. We start AoA estimation from different orientations 40 times. In each experiment, we record the ground truth via a photoelectric encoder and a laser rangefinder.

Figure 5.8 shows the AoA performance in the wide areas. The average errors are 4.5, 6.7, 9.1, 14.2, and 17.7 degrees at 200, 400, 600, 800, and 1000 m, respectively. The errors increase at a long distance. However, the 80% errors are less than 15 degrees when the distance is below 600 m. Since the packet loss becomes unignorable when the distance is more than 800 m, RLoc obtains less efficient ranging difference information in AoA estimation. Nevertheless, the median error is less than 15.3 degrees in the kilometer-level areas and can be exploited for long-range tracking and localization.

5.5.3 NLoS Localization

When the AoA is estimated, we are able to follow the orientation instructions and move to the target node iteratively. Hence, we conduct experiments in a long

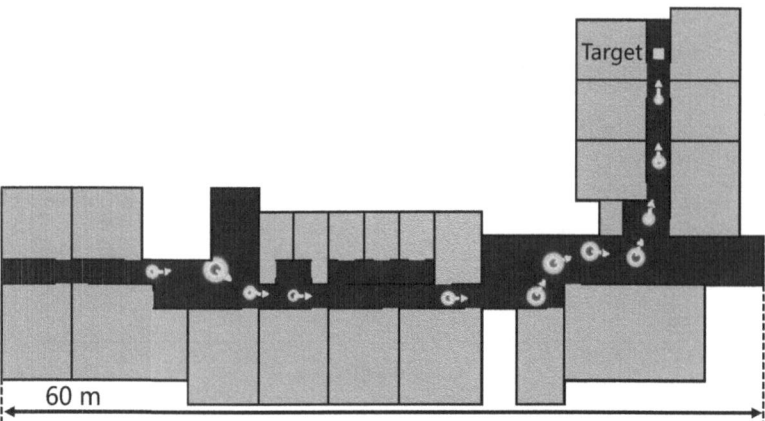

Fig. 5.9 Experimental results on NLoS localization in a long corridor

corridor of a building. Due to the signal attenuation, the UWB and the Bluetooth usually have a limited short localization coverage that is less than 30 m in such scenarios [9, 10]. In the experiments, the target node is fixed at one end of the corridor as illustrated in Fig. 5.9.

In the beginning, the AoA estimation deviates from the actual orientation due to the reflection in the corridor. However, the estimation error becomes smaller as we get closer to the target node. In such scenarios, the transmitted signals have been reflected multiple times before it is received by the gateway. RLoc can successfully follow the AoA of the reflected signals and eventually localize the target node. This also confirms our analysis that AoA estimation does not rely on the accuracy of ToF and it is more stable than ToF in complex indoor environments.

5.5.4   Power Consumption Evaluation

The localization systems designed for IoT applications are sensitive to power consumption since most IoT devices are equipped with a limited battery. In this subsection, we conduct experiments to evaluate the power consumption of RLoc and compare it with the existing localization techniques. We focus on the energy consumption of target nodes since it is costly to recharge them once they are deployed in IoT applications due to large quantities. In the experiment, we record the power for 5 minutes and calculate the average consumption via a power monitor. The comparison results are illustrated in Table 5.2.

In Table 5.2, RLoc is compared with the Wi-PoS (UWB localization) [9], visible light localization [11] and LoRa-based localization [12]. Its power consumption is the minimum among the popular localization systems. In particular, RLoc has a power consumption of 65.8 mW which is only 20% of that consumed by the

Table 5.2 Power consumption comparison

Systems	Technology	Power
Wi-PoS [9]	UWB: ToF	189.0 mW
Pulsar [11]	Visible light: AoA	150.0 mW
Sound-based [12]	LoRa: RSSI	80.0 mW
RLoc	LoRa: AoA&ToF	65.8 mW

UWB localization system (it has a very high localization accuracy with a short coverage range). Since RLoc localizes the target via only anchor, it simplifies the communication and localization that save the power compared to the RSSI based LoRa localization.

5.6 Summary

In this paper, we took the first attempt to enable AoA estimation in LoRa. We proposed to use ranging difference measurements for efficient AoA estimation and designed a tailored ranging procedure. Based on the redesigned gateway, we came up with a novel binary classification framework to achieve a stable AoA estimation. To improve the AoA accuracy, we add more "virtual arrays" by a "rotate and follow" strategy. In RLoc, we focus on static targets in the rotation. When the target is moving, The AoA estimation may be degraded. We are planning to exploit the Doppler shift for the mobile targets since the shift caused by their speed is correlated to the rotation frequency. We consider the 2D AoA estimation and localization in this paper. There are also many applications that require 3D AoA estimation and localization, which has been receiving increasing research attention. We will consider those improvements in future work, by designing a new globular antenna array and applying the idea of AoA estimation for mobile applications. This work provides a new way of AoA estimation and may inspire more work on long-range localization in the future.

References

1. H. Kremo, T. Farrell, J. Tallon, D. McDonald, L. Doyle, A method to enhance ranging resolution for localization of lora sensors, in *Proceedings of the IEEE PIMRC*, 2017
2. Rf Switch hmc241 (2020). https://www.analog.com/media/en/technicaldocumentation/data-sheets/hmc241aqs16e.pdf
3. Sx1280 Lora Calculator (2019). https://www.semtech.com/products/wireless-rf/24-ghz-transceivers/sx1280
4. Application note: an introduction to ranging with the sx1280 transceiver (2017). https://www.semtech.com/products/wireless-rf/24-ghz-transceivers/sx1280

5. K. Qian, C. Wu, Z. Yang, Z. Zhou, X. Wang, Y. Liu, Tuning by turning: enabling phased array signal processing for wifi with inertial sensors, in *Proceedings of the IEEE INFOCOM*, 2016

6. J. Xiong, K. Jamieson, Arraytrack: a fine-grained indoor location system, in *Proceedings of the USENIX NSDI*, 2013

7. W. Gong, J. Liu, Roarray: towards more robust indoor localization using sparse recovery with commodity wifi. IEEE Trans. Mobile Comput. **18**(6), 1380–1392 (2019)

8. M. Kotaru, K. Joshi, D. Bharadia, S. Katti, Spotfi: decimeter level localization using wifi, in *Proceedings of the ACM SIGCOMM*, 2015

9. B. Van Herbruggen, B. Jooris, J. Rossey, M. Ridolf, N. Macoir, Q. Van den Brande, S. Lemey, E. De Poorter, Wi-pos: a low-cost, open source ultra-wideband (uwb) hardware platform with long range sub-ghz backbone. Sensors **19**(7), 1548 (2019)

10. R. Ayyalasomayajula, D. Vasisht, D. Bharadia, Bloc: Csi-based accurate localization for ble tags, in *Proceedings of the ACM CoNEXT*, 2018

11. C. Zhang, X. Zhang, Pulsar: towards ubiquitous visible light localization, in *Proceedings of the ACM MobiCom*, 2017

12. F. Deng, S. Guan, X. Yue, X. Gu, J. Chen, J. Lv, J. Li, Energy-based sound source localization with low power consumption in wireless sensor networks. IEEE Trans. Ind. Electron. **64**(6), 4894–4902 (2017)

Chapter 6
LoRa-Based Indoor Tracking System for Mobile Robots

Abstract The robot's mobility and intelligence have expanded its application in recent years. Specifically, indoor tracking is a fundamental function of public service robots in nursing homes, hospitals, and warehouses. Existing vision-based tracking requires visual information, which may be unavailable and introduce privacy issues in practical deployment. To this end, in this chapter, we propose LTrack, a long-range tracking system based on LoRa, an emerging low-power wide-area networking (LPWAN) technology, with a single transceiver pair. Note that commodity LoRa devices cannot estimate the angle of arrival (AoA) of signals due to hardware limitations. We design a virtual circular antenna array in the mobile rotating anchor via a lightweight hardware modification to multiplex the only RF channel in the low-cost LoRa device. The difference of time of flight (TDoF) measured in the circular antenna array is fused with the rotating orientation to estimate the target AoA. We also redesign and optimize the primitive LoRa ranging engine based on systematic analysis. Further, we present a real-time mobile target tracking algorithm based on the Doppler frequency shift to combat the uncertainty introduced by the target movement. We have developed the prototype of LTrack, which consists of a mobile rotating anchor, a LoRa tag, and a commercial robot. The system is evaluated in both LOS and NOLS indoor scenarios. Experiments show that LTrack supports robust tracking with a median error of 0.12 and 0.45 m in a 137 m^2 lab space and a 600 m^2 corridor, respectively.

Keywords Indoor tracking · Mobile robots · LoRa · AoA estimation

6.1 Estimating the AoA

In this section, we first empower the Commercial Off-The-Shelf (COTS) LoRa chip with AoA estimation capability with a lightweight hardware modification. Then, we optimize the default ranging workflow in the 2.4 GHz LoRa chip to improve the accuracy of AoA estimation.

© The Author(s), under exclusive license to Springer Nature Switzerland AG 2024 75
Z. Shi et al., *LoRa Localization*, SpringerBriefs in Computer Science,
https://doi.org/10.1007/978-3-031-48008-9_6

6.1.1 AoA Estimation with an Antenna Array

As mentioned before, compared with MIMO devices, LoRa transceiver has only one RF channel for uplink and downlink transmissions, and the signal phase information is unavailable due to hardware limitations. Therefore, it is impossible to utilize the phase difference to calculate AoA. To empower LoRa transceiver with AoA estimation capability, we propose to apply a lightweight modification on the COTS LoRa chip, as shown in Fig. 6.1a. The modification requires users to add two antennas and an RF switch, which is easy to implement. The two external antennas are connected to the LoRa analog front end via the RF switch. In this way, we can exploit the difference of the ToFs received at two antennas to calculate the distance difference, denoted by Δd, from the target node to these two antennas, and further calculate the AoA, denoted by θ.

We denote the number of antennas by n, the angle between antenna i and antenna 1 (reference antenna) by α_i, and the radius of the circular antenna array by r. For antenna i and antenna 1, the distance difference, denoted by Δd_{i1}, is

$$\Delta d_{i1} = TDoF_{i1} \times c, i = 2, 3, \cdots, n. \tag{6.1}$$

According to AoA estimation, $\Delta d(\theta)$ is a function of θ and can be expressed as:

$$\Delta d(\theta) = \begin{bmatrix} 2r \sin(\frac{\alpha_1}{2}) \sin(\theta - \frac{\alpha_1}{2}) \\ 2r \sin(\frac{\alpha_2}{2}) \sin(\theta - \frac{\alpha_2}{2}) \\ \cdots \\ 2r \sin(\frac{\alpha_n}{2}) \sin(\theta - \frac{\alpha_n}{2}) \end{bmatrix}. \tag{6.2}$$

From a circular array with n antennas, we can obtain an accurate AoA estimation if Δd measurement is accurate. However, as illustrated in Fig. 6.1b, the LoRa chip needs to process signals received at antennas one by one. In SX1280, it takes about 160 ms to finish a ranging task on 40 subchannels, and the time is doubled in the Time Difference of Flight (TDoF) measurement. During this period, the target may move and the around environment may change, and the AoA θ can not be derived from the TDoF Δd. Thus, it is important to minimize such a waiting time.

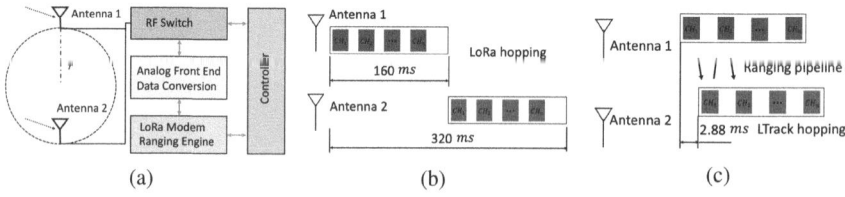

(a)   (b)   (c)

Fig. 6.1 We redesign the standard ranging procedure in 2.4 GHz LoRa. The antennas are selected by the controller via an RF switch, and the hopping procedure is optimized to reduce the time between TDoF estimations from two external antennas. (**a**) Antenna array structure. (**b**) Built-in hopping procedure. (**c**) Optimized hopping procedure

6.1.2 Minimize Ranging Interval

We optimize the hopping process by introducing an antenna switching scheme, in which the anchor selects TX antenna between the hopping process, as shown in Fig. 6.1c. Specifically, the anchor actively controls the RF switch to select the connected antenna before the transmission. When a ranging packet on the CH_1 subchannel is processed from the first antenna (antenna 1 in Fig. 6.1c), we switch to another antenna (antenna 2 in Fig. 6.1c) to transmit a ranging packet on the same subchannel, CH_1, immediately. Then, we switch back to the first antenna for ranging packet transmission on next subchannel, i.e., CH_2. With the antenna switching scheme, the waiting time between two ranging operations on a subchannel decreases from 40 ranging time to 1 ranging time.

Note that an RF switch takes extra 150 ns to select the antenna. We reschedule the ranging process to reduce the interval for each ranging task. In every subchannel, the controller needs to concatenate the ranging parameters, including packet type, frequency, and bandwidth, with the original frame before transmission, increasing processing time. These parameters are usually fixed for tracking a specific target. Hence, we can prepare all the ranging packets in advance when the end device is idle. In particular, the synchronization packet selects the pseudo random frequency hopping channels between the anchor and the target during the ranging. To this end, we move the synchronization packet generation process that was in-between the hopping process ahead. Further, the ranging packets are prepared after the system booting, reducing 13 and 17 commands in the ranging and hopping process, respectively. From our measurements, the proposed reschedule saves 0.218 ms.

After getting a ranging measurement, the LoRa ranging engine calibrates the hardware delay to output final estimation. The hardware delay is design-specific and has been profiled in factory, which can be taken as a constant value. Since we focus on utilizing the TDoFs for AoA estimation, the hardware uncertainty is similar in two adjacent ranging tasks and can be subtracted in the differential data. Thus, instead of performing calibration during the ranging procedure, we move the calibration at the end of AoA estimation to reduce the estimation time. From our measurements, we save 46 ms in an AoA estimation. Finally, the ranging interval is reduced from 160 to 2.88 ms. Hence, the impacts introduced by the target movement is mitigated a lot when calculating the TDoF (a 0.5 m/s speed leads to 0.144 cm location change).

To understand the performance gain of the proposed design quantitatively, we conduct experiments to compare the primitive LoRa ranging procedure and LTrack ranging pipeline regarding to the TDoF estimation accuracy. As shown in Fig. 6.2, the target is placed at the front of the antenna array 3 m away. The mounting points of the two antennas are denoted by A and B, which are separated by 0.2 m. The target is put in the perpendicular bisector of the segment AB, making the ground-truth TDoF be 0. We use the primitive LoRa ranging procedure and LTrack ranging pipeline to measure the TDoF for 250 times. Figure 6.3 presents the histogram of the experiment results of two pipelines. We can see that by using the built-in ranging

Fig. 6.2 The setup of the TDoF mircobenchmark. The target is put in the perpendicular bisector of the segment AB, making the ground-truth TDoF be 0

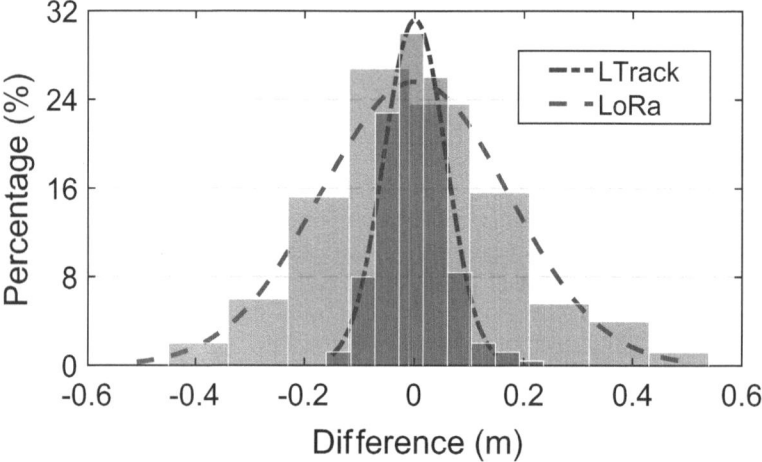

Fig. 6.3 Histogram with density curve of the TDoF measurement results

procedure of LoRa, the values of distance difference Δd varies from -0.5 to 0.5 m. The standard deviation is 0.171 m, which is 85.5% of the distance between antennas. Those fluctuation results in large noise for AoA estimation. The values of Δd vary from -0.2 to 0.2 m when LTrack pipeline is used for TDoF estimation. The standard deviation is 0.057 m, which is 33% of that by using the built-in ranging procedure of LoRa. The results demonstrate that the redesigned ranging procedure is efficient to mitigate the uncertainty in TDoF estimation.

6.2   Eliminating the Blind Area

Although we have enabled the LoRa chip with AoA estimation capability and optimized the primitive ranging pipeline to improve accuracy, the static antenna array cannot remove the impact of the blind area. As shown in Fig. 6.4, in a static antenna array, the antennas block each other and thus generate the blind area. When the target is in the blind area, the AoA estimation accuracy decreases due to the blockage. To overcome this disadvantage, we design a "virtual array" based on the static antenna array proposed in Sect. 6.1.1. As shown in Fig. 6.5, the "virtual array" is a circular antenna array consists of two antennas. The circular antenna array rotates at a certain speed to emulate an antenna array of numerous antennas, and take snapshots of received signals at different spatial locations. During the rotation, the anchor keeps sending and receiving the ranging packet, and records the orientation and speeds of the antenna array. The "virtual array" exploits the spatial diversity of the antennas to eliminate the blind area.

Specifically, the two connected antennas are driven by a DC motor in the anchor. The array orientation $\alpha(t)$ is collected via a photoelectric encoder. Since the TDoF measurements take time of once ranging when the array is rotated, the $\Delta d(t)$ needs to be synchronized with the array orientation. We assume that the first ToF_1 from antenna 1 is measured at time t_1 and the array orientation is $\alpha(t_1)$. When the second ToF_2 from antenna 2 is measured, the array is rotated to angle $\alpha(t_2)$. Therefore, the TDoF calculated from ToF_1 and ToF_2 contains the difference caused by the orientation. The difference caused by the rotation can be calculated by

$$\Delta d_r = 2r \sin(\frac{\alpha(t_2) - \alpha(t_1)}{2}) \sin(\theta - \frac{\alpha(t_1) + \alpha(t_2)}{2}). \tag{6.3}$$

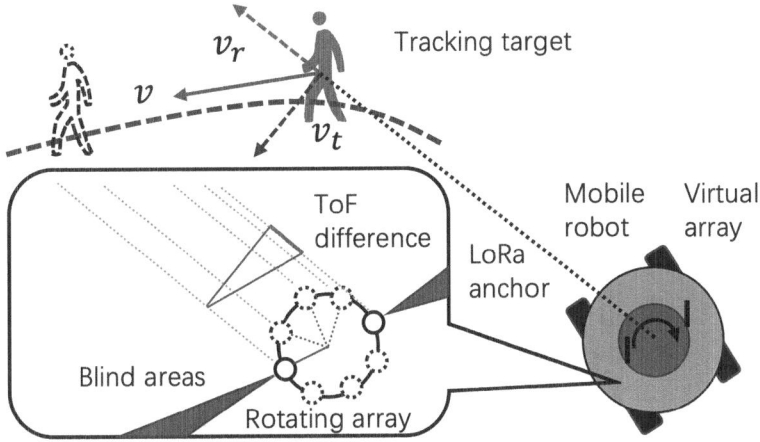

Fig. 6.4 Architecture model

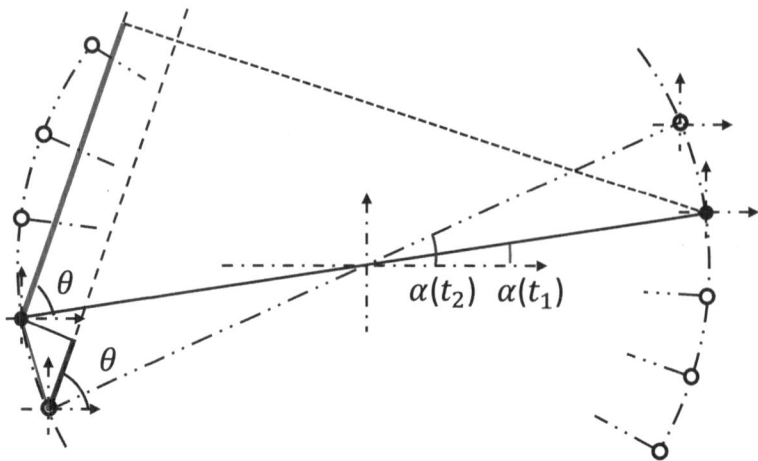

Fig. 6.5 The basic idea of the virtual antenna array. In rotation, the two connected antennas emulate a circular antenna array

Thus, the distance difference can be synchronized by

$$\Delta d = d_2 - d_1 - \Delta d_r. \tag{6.4}$$

In the rotation, the orientation $\alpha(t)$ is recorded at the ranging time t_1, t_2, t_n as $\alpha(t_1), \alpha(t_2), \alpha(t_n)$ and the responding distance difference is $\Delta d_1, \Delta d_2, \dot{A} d_n$. When the anchor collects an array of results, the AoA can be further estimated by

$$\min_{\theta, b} \quad h(\theta, b) = \|\Delta \mathbf{d} - \Delta \mathbf{d}(\theta, b)\|, \tag{6.5}$$

where b is the constant delay caused by the hardware since the calibration is removed in Sect. 6.1.2.

As the object function is not a concave function, this problem is not convex. Thus, it is difficult to calculate the close-form solution. We adopt the SGD optimization [1] to find a solution for the AoA estimation. Specifically, we randomly select a batch comprised of a subset of measured results. The AoA θ and the bias b are initialed and then further updated in the iteration. In each calculation, we select a step factor μ to update the parameter with the gradient of the object function as the follow

$$\begin{bmatrix} \theta_{i+1} \\ b_{i+1} \end{bmatrix} = \begin{bmatrix} \theta_i \\ b_i \end{bmatrix} - \mu \nabla h_{t+1}([\theta_i, b_i]^\mathsf{T}), \tag{6.6}$$

where ∇h_{t+1} is the gradient function. The result are returned when the iteration is finished.

We conduct experiments to compare the AoA estimation accuracy of the static antenna array and the circular antenna array. The tag is placed at a set of points

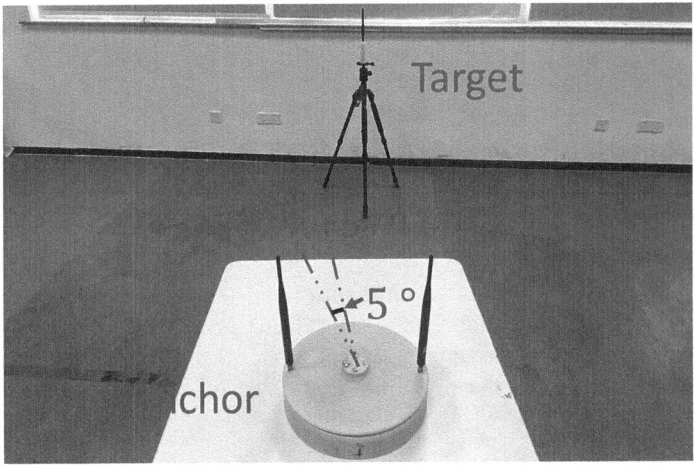

Fig. 6.6 The setup of the AoA estimation accuracy microbenchmark. We place the target at different locations separated by 5 degrees. The distance between the anchor and the target is 3 m

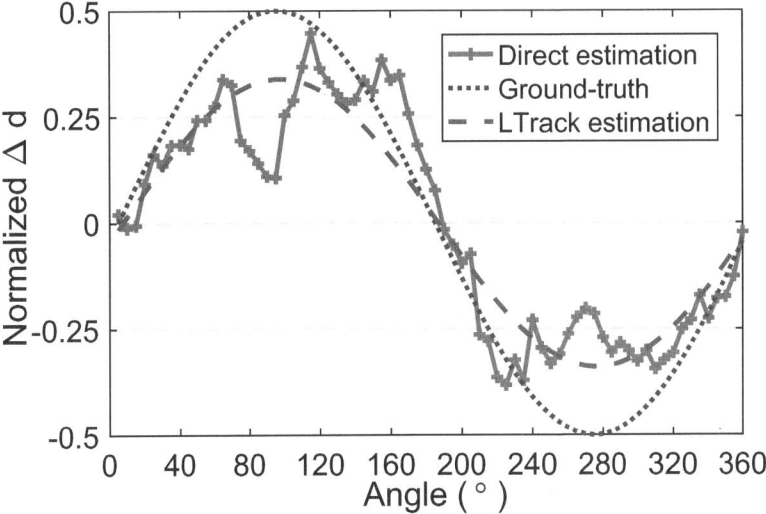

Fig. 6.7 To gain a better understanding, we translate the AoA into TDoF. In the blind areas, direct estimation with a static antenna array bias from the ground truth with an AoA error of more than 40 degrees

around the anchor, and each point is separated by 5 degrees at the distance of 3 m as shown in Fig. 6.6. We translate the AoA estimation results into TDoF to get a better understanding and the measurement results are illustrated in Fig. 6.7. We can see that the error biased from the ground-truth around the 90 degree and 270 degree areas. Only few points achieves accurate result. This is because the noise in two periods are different and even obvious in the blind areas. The fixed antenna array causes the blind area due to the nonlinear model and the obstacles of the antenna itself. Thus, the performance is unrobust in practical environments.

6.3   Estimating Target Movement

Note that SAR has a fundamental assumption, the relative position between the antenna and the target device remains unchanged during the sampling procedure. In our tracking system, the target may move and change its location before the anchor finishing the estimation. Thus, the tracking performance will be degraded. To improve the tracking accuracy, we first model the target motion for moving objects. Then, we implement a real-time frequency estimation algorithm to estimate the target movement.

6.3.1   Target Motion Model

Figure 6.8 illustrates the robot tracks a mobile target in a 2-D space. We can set the start point of the robot as origin to build a coordinate system. The location of the robot at time t is denoted by $[x_r(t), y_r(t)]$. $\varphi_r(t)$ represents the heading orientation of the robot. When the robot tracking the target, it needs adjust its heading orientation from time to time. Thus, the line velocity and angular speed are denoted by $v_r(t)$ and $\omega_r(t)$, respectively. As the anchor is installed on the robot, its location and velocity are equal to the robot's. Note that the circular antenna array rotates at a certain speed, the orientation of the circular antenna array and its rotation speed are denoted by $\alpha_g(t)$ and $\omega_g(t)$, respectively. Similarly, the target location and velocity are denoted by $[x_m(t), y_m(t)]$ and $v_m(t)$, respectively. Specially, the anchor

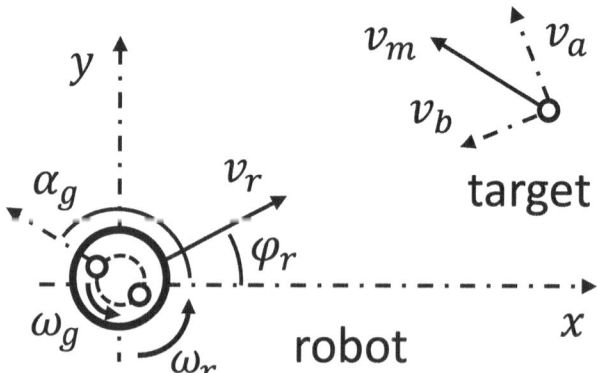

Fig. 6.8 The tracking model considers the tracking in a 2-D space. The anchor estimates the relative distance, angle, and velocity of the target

need to estimate the target relative velocity, as illustrated in Fig. 6.4. We decompose the target velocity v_m into tangential velocity v_a and radial velocity v_b. The radial velocity v_b to the anchor can be calculated from the ranging results. The tangential velocity v_a is proportional to the distance d and the angular velocity ω_m:

$$v_a(t) = \omega_m(t)d(t). \tag{6.7}$$

Note that the observed difference signal frequency ω_o in the anchor is corrected to the antenna rotation speed ω_g and the target angular speed ω_m. According to the Doppler shift model, ω_m can be calculated by

$$\omega_m(t) = \omega_g(t) - \omega_o(t). \tag{6.8}$$

Therefore, we can calculate the target velocity based on the observed signal frequency difference.

6.3.2 Real-Time Frequency Estimation

As the antenna array rotation frequency can be obtained from the anchor, we need to estimate the observed distance difference frequency for real-time tracking. The time-frequency features of non-stationary signals have been widely investigated in the health monitoring, radar systems, Specifically, the continuous wavelet transform (CWT) is a popular method that can calculate the time-frequency features. When the signals $s(t)$ are sampled, the time-frequency coefficients can be obtained via the CWT calculation. As the wavelet scalar is specified as positive, the negative frequency would be lost in the calculation. Thus, we select the anchor frequency that is bigger than the target's velocity. Hence, the coefficients can be calculated as follow

$$F(t, f) = cwt(\Delta d(t)), \tag{6.9}$$

where F is coefficient function of the frequency and time. we further extract the time-frequency features $\omega_o(t)$ via wavelet ridge detection from the coefficients $F(t, f)$.

Based on the frequency estimation, we can calculate the real-time velocity of the target. Thus, we further estimate the real-time AoA of the target via the time-frequency features instead of the motion prediction (assuming the target track in a specific mode with either constant speed or acceleration is not robust for random moving) [2]. When the target track starts from $(\varphi(t_0), d_0)$ and arrive at $(\varphi(t_1), d_1)$, the AoA $\theta(t)$ can be expressed as follow

$$\theta(t) = \theta_0 + \int_{t_0}^{t} \omega_m(t)dt. \tag{6.10}$$

Thus, the distance vector can be modeled with the time-frequency features and expressed as

$$\Delta d(\theta(t)) = \begin{bmatrix} 2r \sin(\frac{\alpha_{t_1}}{2}) \sin(\theta(t_1) - \frac{\alpha_{t_1}}{2}) \\ 2r \sin(\frac{\alpha_{t_2}}{2}) \sin(\theta(t_2) - \frac{\alpha_{t_2}}{2}) \\ \cdots \\ 2r \sin(\frac{\alpha_{t_n}}{2}) \sin(\theta(t_n) - \frac{\alpha_{t_n}}{2}) \end{bmatrix}. \tag{6.11}$$

Correspondingly, the real-time AoA is estimated by

$$\min_{\theta_{t_0},b} \quad h(\theta_{t_0}, b) = \|\Delta d - \Delta d(\theta(t), b)\|. \tag{6.12}$$

Obviously, this objective function optimization is similar to the static AoA calculation. Thus, we estimate $\theta(t)$ based on the iterative method and design the tracking algorithm as illustrated in Algorithm 3. To track a moving target, the

---

Algorithm 3: Tracking optimization

Input: maximum times of iteration $iter_{max}$, iterative threshold τ, the step factor μ, the distance threshold d.
Output: the target node location $\tilde{p}$
Start rotating the antenna array and record its orientation $\alpha_g(t)$, its rotation speed $\omega_g(t)$, the robot location p_r, the robot orientation φ_r, and the robot rotation speed ω_r;
Initialize iterative error $e_{iter} = \tau$, $\theta = \theta_0$, $b = b_0$;
while $d_t \leq d$ **do**
    Initialize $iter = 0$;
    collect new ranging result $\Delta d(t)$;
    Calculate $\theta(t)$ via (6.9) and (6.10);
    while $iter \leq iter_{max}$ **do**
        Select m samples $\theta_m \Delta d_m$ random from the results;
        Calculate $h_t = h(\theta, b)$ via (6.12);
        Update θ and b via (6.6);
        if $h_t - h_{t+1} \leq \tau$ **then**
            $\theta = \theta_{t+1}$;
            $b = b_{t+1}$;
            break ;
        end
        $h_{t+1} = h_t$;
        $iter = iter + 1$;
    end
    Calculate the target relative location $\tilde{p}$ via (θ, d) and move to this destination;
end
return $\tilde{p}$;

---

robot estimates its relative real-time location and schedules the path to move to the destination. Specifically, the mobile anchor keeps sending the ranging requests and measuring the rotation angle in the tracking. The ranging difference $\Delta d(t)$ and the corresponding angle $\varphi(t)$ are recorded in a measurement queue. As the target may moves in the tracking, the target speeds are first estimated via the Doppler shift. Further, the real-time AoA is calculated in the stochastic gradient descent optimization, during which the m random samples from the results are selected in each iteration until the error is less than the threshold or the iteration is finished. When the real-time AoA is calculated, the robot selects a reachable path to follow the moving target's location.

6.4 System Implementation

As shown in Fig. 6.9, we implement a prototype of LTrack consists of three components, including a tag, an anchor, and a firmware for the commercial mobile robot. The tag can be attached to static or moving objects. The anchor contains a circular antenna array that is installed on the robot. The architecture of the firmware is shown in Fig. 6.9b. When the anchor needs to track the target, the result is sent to the robot controller via a wireless serial port for the target tracking. We developed the tracking firmware on the robot controller platform (Linux-armv7hf) to listen for the ranging result and calculate the destination location. The robot keeps tracking and selects a new path to move to the target when the result is updated.

Tag The LTrack tag needs to be attached to the tracking objects and only sizes 4 cm × 10 cm because it only has essential components for communication and tracking. We simplify the hardware design to save power and extend battery life. The tag uses an ultra-low-power microcontroller STM32L476 and an SX1280 LoRa chip. The tag transmits and receives messages or responses to ranging packets sent by the anchor. Note that our design is compatible with the standard frequency hopping protocol. Thus, any devices that support 2.4 GHz LoRa ranging can serve as the tag. We choose SX1280 to prototype our system due to its wide availability. The cost of a LTrack tag is about 12 US$.

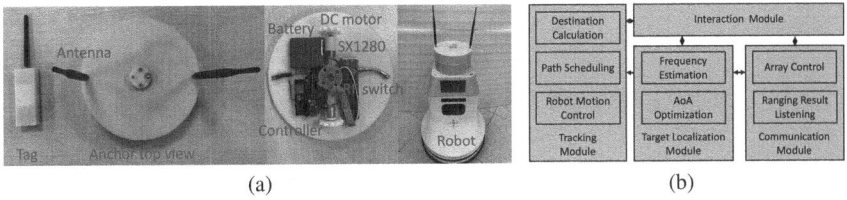

(a)   (b)

Fig. 6.9 Hardware implementation and firmware design of the LTrack system on a robot. (**a**) Hardware implementation. (**b**) Firmware framework

Anchor The LTrack anchor runs the proposed ranging tasks and coordinates with the robot through wireless serial ports. The anchor is equipped with a circular antenna array with two omnidirectional antennas. Follow the design in Sect. 6.1.1, the two antennas are separated by 0.2 m. We connect the two antennas to a HMC241 RF switch to multiplex the single RF channel of the SX1280 LoRa chip. We install those components on a rotary board that is driven by a DC motor. The angle of rotation is recorded via a photoelectric rotary encoder. The speed of rotation is controlled by a Proportional Integral Differential (PID) controller that drives the DC motor with pulse-width modulation. The cost of a LTrack anchor is about 20 US$.

Robot Firmware To apply the LTrack system on a real robot, we develop a new robot firmware. As illustrated in Fig. 6.9b, the firmware has a communication module, a target localization module, a tracking module, and an interaction module. The communication module bridges the LTrack anchor and the robot controller via wireless serial ports. The target location module performs real-time AoA and frequency estimations and sends the results along with the ToF estimation results to the tracking model. The tracking module fuses the results and schedules the tracking path. The interaction process the voice command from the user to identify which LTrack tag to track. In our settings, the firmware is running on a Khadas VIM3 single-board computer, which belongs to the commercial robot.

6.5 Performance Evaluation

In this section, we conduct practical experiments in complex indoor environments to evaluate the performance of LTrack system. We also demonstrate the deployability of apply our system in a large indoor space through simulations.

6.5.1 Indoor Experiments

We first evaluate the AoA estimation accuracy in LOS and NLOS scenarios, and then the overall tracking performance in different environments with the prototype is described in Sect. 6.4.

6.5.1.1 LOS AoA Estimation

Setup The bandwidth and the spreading factor of the LoRa are set to 1625 kHz and SF5, respectively. The transmitting power of the tag and the anchor is 12.5 dBm. The rotation speed of the circular antenna array is 40 degree/s. We place the LTrack tag away from the LTrack anchor at different distances, i.e, 40, 50, and 60 m. As

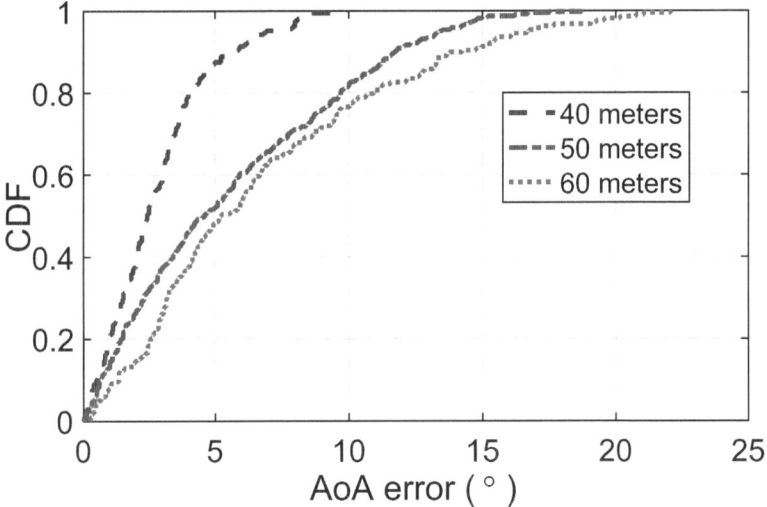

Fig. 6.10 LOS AoA estimation errors in different ranges

shown in Fig. 6.11, the anchor is placed on a tripod at the same height of the target and starts the estimation from different orientation. For each distance setting, we let the anchor estimate the AoA of the tag 200 times.

Results Figure 6.10 shows the CDF of AoA estimation error when tracking the tag located at different distances with LOS. The LTrack anchor is able to track the tag with a median error of 2.4 degrees when they are 40 m apart from each other. The LTrack achieve AoA estimation errors 4.5 degrees at 50 m, and 5.2 degrees at 60 m. We can see that AoA estimation error increases with distance. This is because the signal attenuates after the long distance propagation. We also notice that the estimation accuracy are similar at 50 and 60 m, where their median errors differ by 0.7 degrees. In particular, near 85% of estimation errors are less than 5 degrees at 40 m. The 5 degrees AoA error can be translated into a 3.5 m ($40 \text{ m} * \tan 5°$) distance from the actual location of the tag. The performance will improve as the robot moves towards the target, and thus has a shorter distance.

6.5.1.2 NLOS AoA Estimation

Setup When the robot is tracking an object, the signal path between the anchor and the tag may be NLOS. Thus, we place the tag behind the obstacle to block the signal path. As shown in Fig. 6.11, we use three kinds of obstacles to create NLOS scenarios, including a foam board, a wooden table, and a concrete pillar. The distance between the anchor and the tag is 50 m. For each type of the obstacle, we

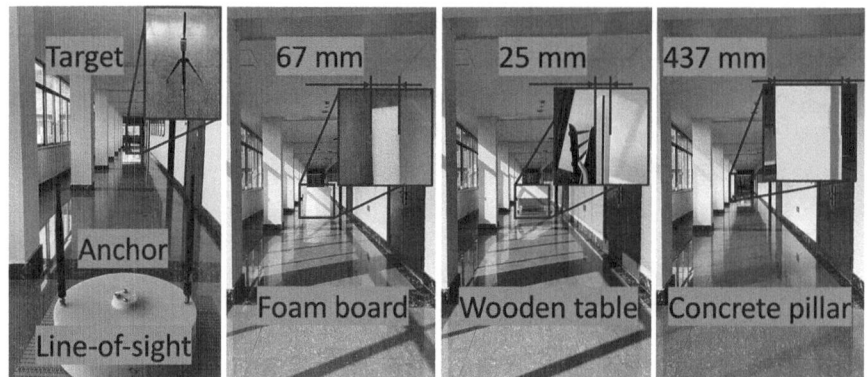

Fig. 6.11 LOS and NLOS AoA estimations

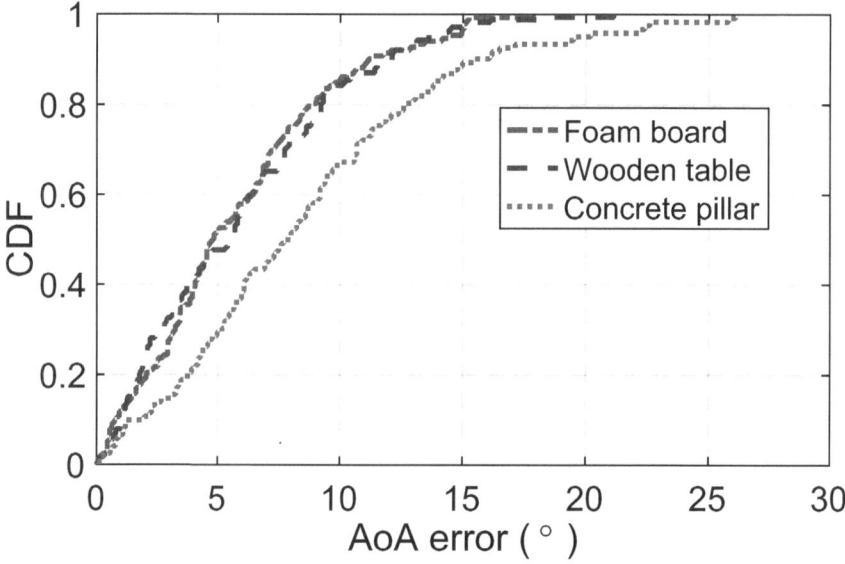

Fig. 6.12 NLOS AoA estimation error with blockage of different obstacles 50 m away from the anchor

perform the AoA estimation for 200 times. The parameters of LoRa radio are same with Sect. 6.5.1.1.

Results Figure 6.12 presents the CDF of AoA estimation error when different types of obstacles block the signal path. LTrack respectively achieves a median AoA estimation error of 5.72 degrees, 5.83 degrees, and 8.49 degrees, with the blockage of the foam board, wooden table, and concrete pillar. Compared with the LOS scenario, the performance of AoA estimation degrades in NLOS settings.

The NLOS results meet our expectations because the signal attenuates significantly due to reflections, scattering, and diffraction. We find that the foam and wooden obstacles have fewer negative impacts on the AoA estimation performance because the LoRa signal can easily penetrate these obstacles. Nevertheless, the performance degradation caused by NLOS paths can be mitigated by the fact that the direct signal path between the anchor and the tag changes with the robot's movement.

6.5.1.3 Tracking in a Lab Space

Setup We evaluate the tracking performance of LTrack in a $137\,\text{m}^2$ lab space. Figure 6.13 shows the floor plan of the lab. The lab has a smart manufacturing testbed with many metal devices (e.g., robot arms) and many desks with PC or personal staffs on them, as shown in Fig. 6.14a and b, respectively. The user holds the LTrack tag in hand and walks naturally along a preset path, which has been

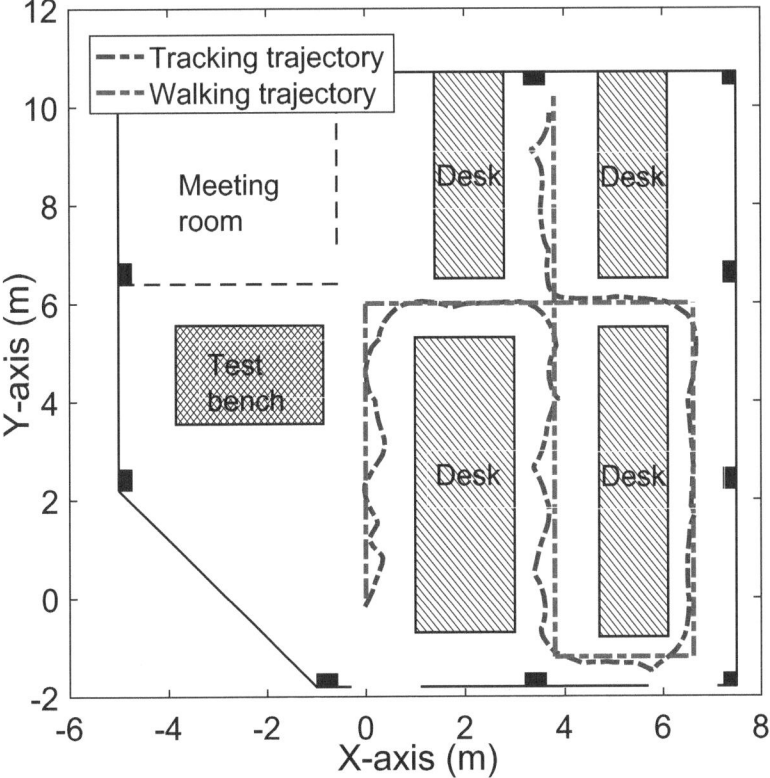

Fig. 6.13 The tracking experiment in an indoor lab space. The robot begins tracking at $(0, 0)$ and finishes at $(3.8, 10)$

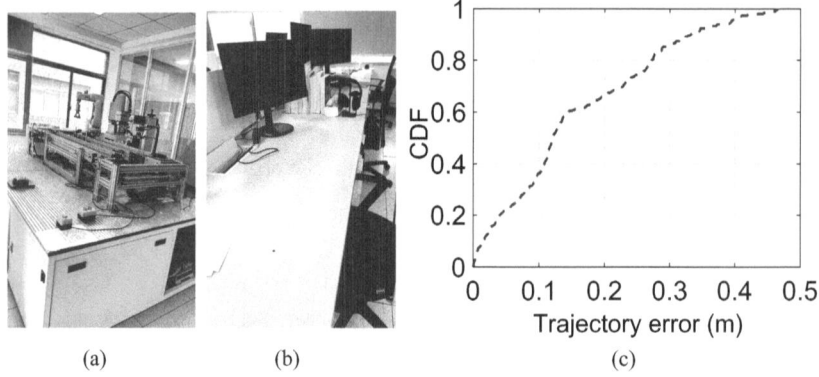

Fig. 6.14 Obstacles in the lab and the tracking error. (**a**) Test bench. (**b**) Desk. (**c**) CDF of the tracking error

marked in red in Fig. 6.13. The anchor and the robot estimate the tag location to follow the user. The parameters of LoRa radio are consistent with Sect. 6.5.1.1.

Results The dashed blue line in Fig. 6.13 represents the tracking trajectory of the robot. We compute the tracking error according to the euclidean distance between the robot's tracking trajectory and the ground truth. Figure 6.14c presents the CDF of the tracking error. The LTrack system achieves a median error of 0.12 m and an 80 percentile error of 0.27 m. During the tracking, we set a safe distance (i.e., 1–2 m) between the target and the robot as the robot may deviate from the ground-truth trajectory.

6.5.1.4 Tracking in a Corridor

Setup We further conduct tracking experiments in a corridor sized $10 \times 60\,\text{m}^2$. Figure 6.15 shows the floor plan of the corridor. There are many concrete pillars in the corridor. We adopt the same setting in Sect. 6.5.1.3, except the user walking speed. The user walks along the corridor at speeds of 0.3 and 0.5 m/s.

Result Figures 6.15 and 6.16 illustrate the tracking results at different user walking speeds. Figure 6.17 presents the CDFs of the tracking errors. We can see that the tracking error increases when the user walks at a faster speed. LTrack estimates the tag's velocity based on discretely collected samples at a certain frequency instead of continuously sampling. The discrete sampling cannot capture some motion features when the tracking target moves rapidly. The LTrack system still achieves sub-meter tracking accuracy. The median errors at the speed of 0.3 m/s and 0.5 m/s are 0.24 m and 0.45 m, respectively.

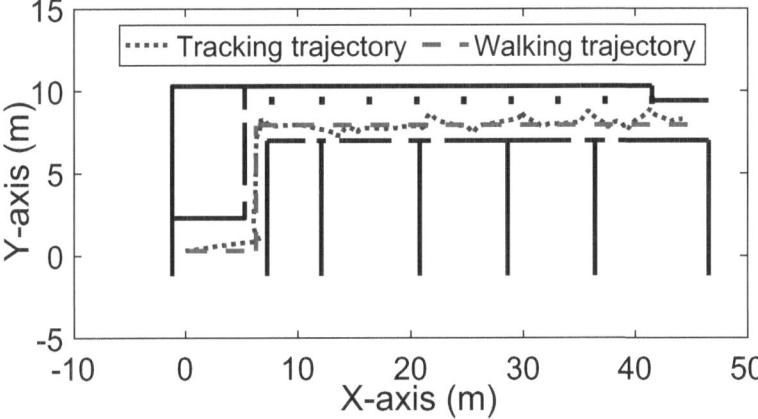

Fig. 6.15 Tracking experiment in a corridor. A person holds the tag in nature and walks at a speed of 0.3 m/s

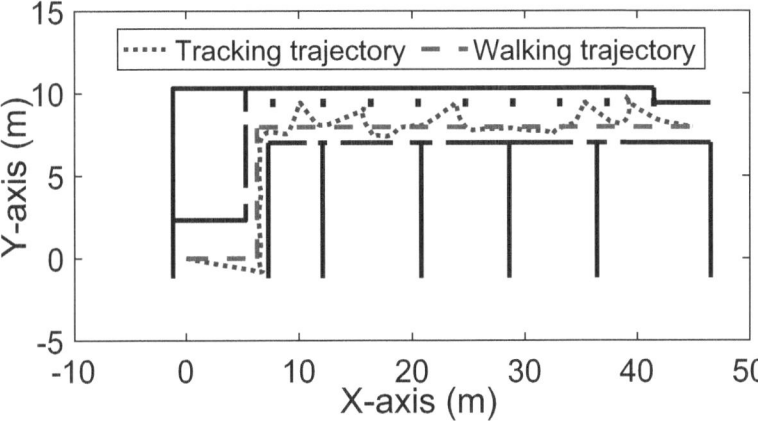

Fig. 6.16 Tracking experiment in a corridor. A person holds the tag in nature and walks at a speed of 0.5 m/s

6.5.2   *Deploy Ability Investigation*

Based on our system model and data traces retrieved from experiments in Sect. 6.5.1, we investigate the deployability of the **LTrack** via simulations. We model the signal quality based on real RSSI measurements at different distances. The path-loss model is fitted as $\text{RSSI}_d = -28.31 * log10(d) - 26.96$. Further, the SNR is derived from the RSSI as $\text{SNR} = \text{RSSI} - (-90)$. The antenna rotation speed and the ranging period are the same as experiments before.

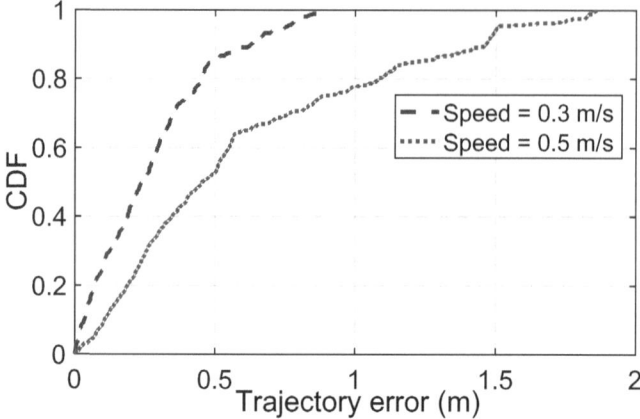

Fig. 6.17 Tracking errors with different walking speeds

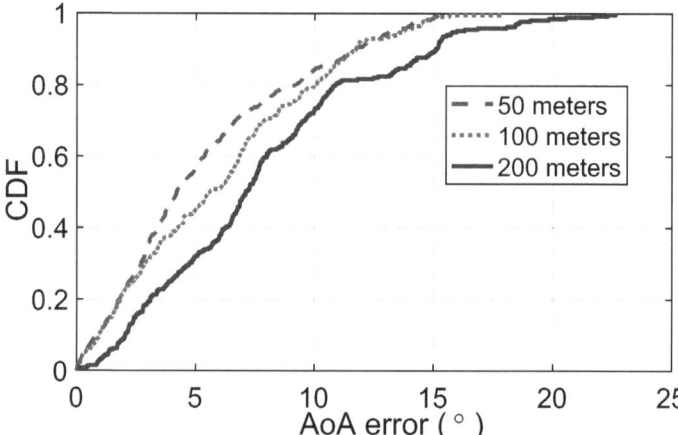

Fig. 6.18 LTrack AoA estimation performance at different ranges

Distance The LoRa signal attenuates with distance, which impacts the performance of the AoA estimation. To understand how distance affects the AoA estimation accuracy, we run simulations at different distances. The tag is apart from the anchor with the distance of 50 m, 100 m, and 200 m. For each distance, we collect 200 estimation values, and the CDF is illustrated in Fig. 6.18. The median AoA estimation errors in 50, 100, and 200 m are 4.3 degree, 5.6 degree, and 7.1 degree, respectively. We find that the simulation results at 50 m well matches our experimental results. In the path-loss model, the signal attenuates slowly when the distance increases. For example, the RSSI decreases 22 dBm when the distance changes from 10 to 60 m. But it just attenuates 5 dBm when the distance changes from 100 to 150 m. Thus, even the distance is 200 m, the AoA estimation median error is still less than 10 degree, which

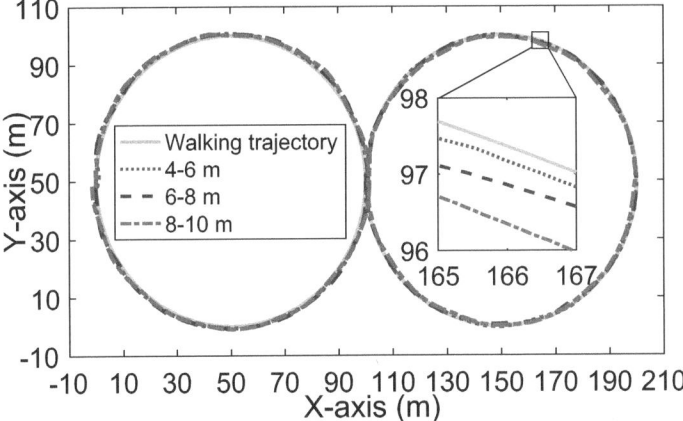

Fig. 6.19 Tracking simulation in a large indoor space with different safe distance settings

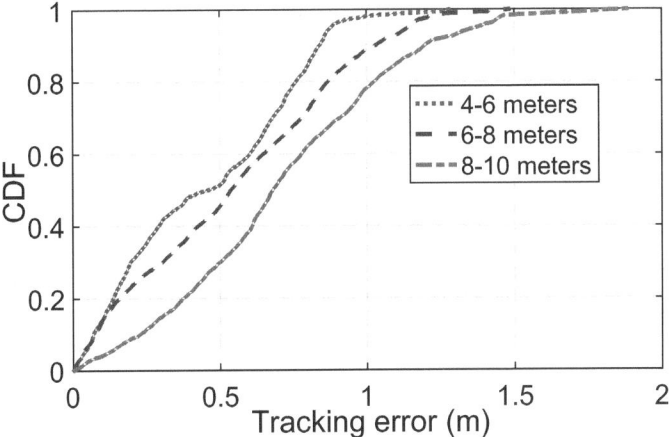

Fig. 6.20 Tracking errors with different safe distance settings

is sufficient for the robot to move close to the target. As the robot is approaching the target, the AoA estimation accuracy increases and will be stable.

Scalability To investigate the scalability of the LTrack system, we extend the tracking area to a $100 \times 200\,\text{m}^2$ space. As shown in Fig. 6.19, the target starts from the point (0, 50) with a 1 m/s moving speed. The robot may keep a safe distance between the target and itself. Thus, we let the robot track the target with a safe distance of 4–6, 6–8, and 8–10 m in three settings. Figure 6.20 presents the CDF of the tracking error. The median errors of tracking distances of 4–6 m, 6–8 m, and 8–10 m are 0.45 m, 0.55 m, and 0.72 m, respectively. Although the safe distance impacts the performance, the LTrack system still achieves an 80 percentile error of

1.03 m in the 20,000 m^2 indoor space. These results indicates that the LTrack system is feasible to track a target with different safe distance requirements in a large indoor space.

6.6   Summary

This paper presents LTrack, a system that allows mobile robots to perform indoor tracking using 2.4 GHz LoRa signals without prior deployed infrastructure. LTrack enables LoRa devices to estimate AoA and track moving objects by a set of hardware and software designs. We develop an LTrack prototype system on a mobile robot. Experiments show that LTrack can track a moving object with decimeters level accuracy. LTrack offers an infrastructure-free, low-cost, and lightweight approach for mobile robots to track objects in the indoor environment.

References

1. L. Bottou, Large-scale machine learning with stochastic gradient descent, in *Proceedings of COMPSTAT'2010* (Springer, Berlin, 2010), pp. 177–186
2. C. Li, Z. Shi, J. Chen, Hardware architecture and optimisation of FPP particle PHD filter for multi-target tracking in cyber-physical systems. IET Control Theory Appl. **11**(11), 1830–1837 (2017)

Chapter 7
Conclusion and Future Directions

Abstract In this chapter, we summarize the presented research and discuss some future directions for low-power IoT nodes localization.

Keywords Internet of Things · LoRa · Localization · Power · Scalability · Artificial Intelligence

7.1 Concluding Remarks

Throughout this monograph, we have delved into the research and development of low-power IoT node positioning techniques, with a specific focus on leveraging the capabilities of LoRa technology. We have explored various aspects of low-power IoT, including network architectures, technical characteristics, and industrial applications. By classifying and summarizing existing research on node positioning, we identified the associated deficiencies and challenges. Building upon this foundation, we presented different positioning models, hardware platforms, and algorithms to achieve accurate and efficient node localization in both indoor and outdoor environments.

The research presented in this book has made significant contributions to the field of low-power IoT node positioning. We have introduced innovative approaches and demonstrated their effectiveness through extensive experimental results. The achievements can be summarized as follows:

- In Chap. 3, we designed a node location algorithm based on the multilateral positioning principle, optimizing the gateway's moving trajectory and integrating sensor data to improve positioning accuracy and reduce power consumption. The experimental results showcased the system's impressive performance in outdoor open environments.
- Chapter 4 presented a low-cost and low-power outdoor positioning system based on LoRa Mesh networking. By leveraging multi-anchor wireless Mesh networking and multi-dimensional data fusion, the system achieved high-precision positioning and wide-area coverage. The results demonstrated the system's

capability to meet the outdoor positioning requirements of low-power and low-cost devices.

- In Chap. 5, we addressed the challenges of indoor positioning in complex environments by introducing a localization method based on signal arrival angles. By utilizing signal flight time differences and designing an antenna array structure, we achieved accurate angle estimation and enabled node localization in complex indoor environments.
- Chapter 6 investigated fusion localization and tracking based on mobile robots, integrating indoor and outdoor positioning technologies. By dynamically adjusting node positioning model parameters and analyzing the time-frequency characteristics of LoRa signals, the system achieved accurate fusion positioning and trajectory estimation in various indoor and outdoor environments.

The research findings and experimental results presented in this monograph demonstrate the potential and effectiveness of low-power IoT node positioning using LoRa technology. By addressing the challenges and deficiencies in existing research, we have contributed to advancing the field and opening up new opportunities for practical applications.

7.2   Future Directions

While this monograph has made significant strides in low-power IoT node positioning, there are still several avenues for future research and development. We identify the following areas as promising directions for further exploration:

Enhancing Localization Accuracy Despite the achieved accuracy in various scenarios, there is room for further improvement. Future research should focus on developing advanced algorithms and techniques to enhance the accuracy of node localization in both indoor and outdoor environments. Additionally, investigating the fusion of multiple positioning technologies and data sources can lead to more robust and accurate localization solutions.

Power Optimization and Energy Harvesting Power consumption remains a critical aspect in low-power IoT node positioning. Future studies should explore innovative techniques to optimize power consumption and extend the battery life of IoT devices. Additionally, the integration of energy harvesting techniques, such as solar or kinetic energy, can provide sustainable power sources for IoT nodes.

Deployment Scalability and Network Optimization As the scale of IoT deployments continues to grow, it becomes essential to develop scalable and efficient network optimization algorithms. Future research should focus on designing algorithms and protocols that can handle large-scale IoT deployments while minimizing network congestion, optimizing communication efficiency, and ensuring reliable positioning performance.

Integration with Edge Computing and AI The integration of edge computing and Artificial Intelligence (AI) can further enhance the capabilities of low-power IoT node positioning. By leveraging edge computing resources, real-time data processing, and distributed decision-making can be achieved, enabling more intelligent and context-aware positioning systems.